[英国]马克·马斯林 著　朱邦芊 译

牛津通识读本·

气候

Climate

A Very Short Introduction

译林出版社

图书在版编目（CIP）数据

气候 ／（英）马克·马斯林（Mark Maslin）著；
朱邦芊译． —南京：译林出版社，2022.8
（牛津通识读本）
书名原文：Climate：A Very Short Introduction
ISBN 978-7-5447-9181-6

I.①气⋯　II.①马⋯ ②朱⋯　III.①气候学－普及
读物　IV.①P46-49

中国版本图书馆 CIP 数据核字（2022）第 083874 号

著作权合同登记号　图字：10-2017-080 号

气候　[英国] 马克·马斯林／著　朱邦芊／译

责任编辑　杨欣露
装帧设计　景秋萍
校　　对　王　敏
责任印制　董　虎

原文出版　Oxford University Press, 2013
出版发行　译林出版社
地　　址　南京市湖南路 1 号 A 楼
邮　　箱　yilin@yilin.com
网　　址　www.yilin.com
市场热线　025-86633278
排　　版　南京展望文化发展有限公司
印　　刷　江苏扬中印刷有限公司
开　　本　890 毫米 ×1260 毫米　1/32
印　　张　10.125
插　　页　4
版　　次　2022 年 8 月第 1 版
印　　次　2022 年 8 月第 1 次印刷
书　　号　ISBN 978-7-5447-9181-6
定　　价　39.00 元

序 言

谈哲敏

译林出版社推出了一套"牛津通识读本"(Very Short Introductions),邀请我为其中的《气候》一书作序。当今气候变化问题已不仅仅是一个自然科学问题,还涉及政治、经济、法律等诸多方面,甚至与人类健康、地球的未来息息相关,需要引起公众的高度关注,因此我欣然应允为本书作序。这本七万余字的小书简洁精练,两三天的零散时间足够通读。该书作者马克·马斯林毕业于剑桥大学,目前是伦敦大学学院的地球系统科学教授,研究领域包括过去和未来全球气候变化的原因及其对全球碳循环、生物多样性、热带雨林和人类进化的影响。马斯林先生曾在包括《科学》《自然》《柳叶刀》等期刊上发表了170余篇论文,还以气候变化等为主题参与撰写了十多本著作。他常年致力于向公众传播气候、生态和环保领域的前沿知识。由这样一位资深且专业的学者撰写本书,自然再合适不过。

在本书中,马斯林先生站在地球系统科学的角度带领读者认识气候。他使用浅显易懂的语言和丰富的图表,向读者介绍了大气科学中一系列重要的基本概念与基本现象，如温室效应、大气环流、混沌理论、恩索（ENSO）、季风等，并介绍了气候系统与海洋、与板块运动之间的相互关联。作者同时也站在时间长河中观察气候变化，介绍了过去亿万年时间尺度上的气候变化趋势及典型事件。马斯林先生这种空间和时间上大跨度的论述视角，有助于读者建立起气候系统牵一发而动全身的宏观认识，全面、动态地看待气候系统，这一独特的写作方式，与本书作为"通识读本"的定位十分契合。

马斯林先生也展示出其作为一名科学家的社会责任感。他介绍了气候预测的概念，并提出不断改进的气候模型能够帮助世界各国进一步提高灾害性天气气候事件的监测、预报、预警能力，为制定防灾减灾、大气污染的治理决策提供科学支撑。但作者也流露出深切的忧患意识，指出全球变暖背景下，气候预测将变得更加困难。未来的气候变化与减贫、环境恶化和全球安全一样，是21世纪的决定性挑战之一，风暴、洪水、热浪和干旱等极端气候事件的发生将变得更加频繁。作者列举众多科学研究证据，有力证明了人类活动对气候变化造成的巨大影响，并据此给出了应对气候变化的一些主要方法。在本书中，作者也赞扬了我国近年来通过大力推行植树造林实现生物除碳的举措。在碳达峰目标、碳中和愿景下，我国实施积极应对气候变化国家战略，必将为全球应对气候变化贡献重要力量。

恩格斯在《自然辩证法》中曾明确指出："我们不要过分陶醉于我们人类对自然界的胜利。对于每一次这样的胜利，自然界都将对我们进行报复。"了解气候系统的知识是应对复杂的气候变化问题的基础。这本以气候为主题的小书从科学的角度展现了气候系统令人惊叹的一面，帮助我们认识世界的神奇和美丽，增加对世界的另一层欣赏。作为一本中英双语对照的科普读物，本书不仅适合大气、地理、生态、地球科学等专业的师生学习，更值得每一位关心地球环境的读者阅读。

目　录

致　谢

作者想要感谢安妮、克里斯、约翰娜、亚历山德拉、阿比·马斯林，以及休·安德鲁斯的陪伴；感谢埃玛·马钱特和莱萨·梅农出色的编辑与支持；还要感谢伦敦大学学院环境研究所、伦敦大学学院地理系、TippingPoint 社区、Rezatec 有限公司、二叠纪公司、DMCii 公司、KMatrix 公司，以及全球贵重商品公开有限责任公司的全体员工和朋友们；最后，感谢迈尔斯·欧文绘制的精彩插图，它们同样重要。

i

第一章

气候是什么？

引 言

　　从今天穿什么衣服到最近染上了什么病，气候影响着我们生活中的一切，因为我们人类只有在一个非常狭窄的温湿度范围内才会感到舒适。这个舒适区间的温度范围大约为20 ～ 26 ℃，相对湿度在20% ～ 75%之间（图1）。然而，我们几乎可以在世界的每一个角落生存，也就是说生存条件往往会在这个舒适区间之外，而我们也学会了通过增减衣物、改造住房来保持舒适。因此，你也许会认为自己的衣柜里挂着的衣服只是反映了你有没有时尚品位，但实际上，它们展现了你生活所在地的气候及其一年四季的变化。所以你有一件厚厚的加绒大衣以备在加拿大过冬，也有一件短袖衬衫应付里约热内卢的商务会议。衣柜也暗示了我们喜欢去哪里度假——崭露头角的极地探险家的衣柜里会挂着非常保暖的北极服装，如果喜欢在沙滩上晒太阳，那么衣柜里就会改挂短裤或比基尼。

房子也是在我们清楚了解本地气候的情况下建造的。在英格兰，因为室外的温度通常低于20 ℃，所以几乎所有的房子都有中央供暖，同时由于温度很少超过26 ℃，因此又鲜有空调。然而在澳大利亚，大多数房子都有空调，却很少有中央供暖。气候还会影响城市的结构，以及世界各地运输系统的运作方式。

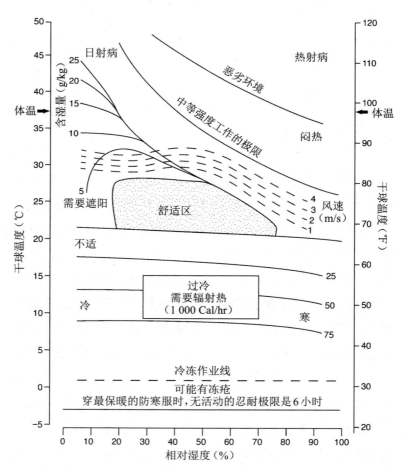

图1　人类的生活舒适情况与气候

得克萨斯州休斯顿有长达7英里的地下隧道网络，连接着市中心所有的主要建筑；这个全气候控制的网络把95个人口密集的城市街区连在一处。当外面下雨或天气炎热的时候，人们都走隧道，因为休斯顿每年至少有五个月的平均气温高于30 ℃。同样，加拿大也有地下商场，以此来避免大雪和极寒的问题。

气候决定了我们在何时何地获得食物，因为农作物生产既受霜冻雨雪的限制，也被生长季节的长短所左右，其中包括日照量和温暖季节的长度。因此简而言之，水稻生长在温暖和非常潮湿的地方，而小麦可以在范围大得多的温带气候区生长。气候也会影响食物的质量，例如，众所周知，法国葡萄酒的最佳酿造年份需要冬季有几次短暂的严寒，可以让葡萄藤更加强壮从而产出优质的葡萄。农民也可以"改善"当地的气候，比如在温室里种植番茄，或是通过灌溉土地来提供更稳定的供水。

气候还会影响有热浪、干旱、洪水和风暴等极端天气事件的地方。但在很多情况下，我们对极端事件的认识取决于当地的条件，例如，2003年北欧遭遇了"热浪"，英格兰首次出现了100 °F（37.8 ℃）的高温纪录。然而在热带国家，只有当气温超过45 ℃时才会作为热浪记录下来。气候对我们的健康也有很大的影响，因为很多疾病都受到温湿度的制约，例如流行性感冒（俗称流感）的发病率会在冬季达到高峰。由于南北半球的冬季时间不同，全球每年实际上会有两个不同的流感季。每个冬季过后，流感病毒都会在两个半球之间迁徙，让我们有时间根据另一个半球半年前出现的新型毒株来生产新的疫苗。关于流感为何受气候控制的说法有很多，从理论上讲，在寒冷干燥的条件下，病毒在各种物体表面存活的时间更长，所以更容易在人与人之间传播。另一种说

法是,维生素D可能会帮助人们对病毒产生一定的抵抗力或免疫

3 力。因此,在冬季和热带的雨季,当人们待在室内、远离阳光时,他们体内的维生素D水平会下降,流感的发病率就会上升。

冷暖地球

地球上气候的成因是赤道所接收的太阳能量多于两极。如果把地球想象成一个巨大的球体,离太阳最近的点就是赤道或球体中部。赤道是太阳照射最直接的地方,因而这里也是地球接收能量最多之处。离开赤道向北或向南,地球表面沿曲线远离太阳,地表与太阳所成夹角增大。这意味着太阳的能量被分散在更大的区域,从而导致温度降低。如果我们生活在一个

4 平坦的圆盘上,就会从太阳那里得到更多的能量——大约每平方米1 370瓦特(1 370 W/m²),而地表由于其弯曲的性质,获得的能量为平均每平方米343瓦特左右(图2)。在太阳迸发出的能量中,地球只接收了非常少的一部分。只需想想太阳每释放20亿瓦特的能量,我们只接收到1瓦特,就知道地球与太阳相比有多渺小了。这也是为什么在很多科幻小说中,作者都会想象恒星被条带甚或曲面所环绕,借此收集直接流失到太空的全部能量。

在我们接收到的太阳能中,大约有三分之一都被直接反射回太空了。这是由"反照率",即表面反射率的大小,所造成的。例如白云和白雪的反照率很高,几乎可以把照在其上的阳光全部反射回去,而海洋、草地和雨林等比较暗的表面则会吸收更多的能量。两极接收的能量不仅比赤道少,而且还向太空散失了

5 更多的能量(图3):南北两极白色冰雪的反照率很高,把大量的

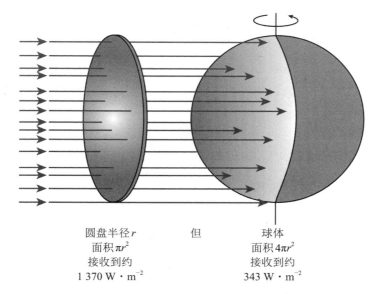

圆盘半径 r   但   球体
面积 πr²   面积 4πr²
接收到约   接收到约
1 370 W·m⁻²   343 W·m⁻²

图2 太阳能在球面上的分布

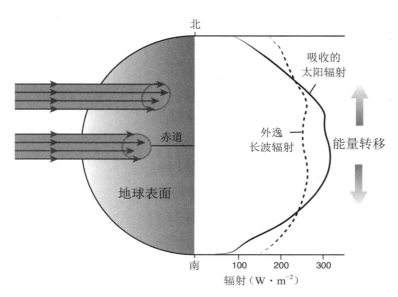

图3 在太阳辐射角度的驱动下,能量从赤道转移

5

太阳能反射回了太空。与此相反，颜色暗得多、反射率也更低的低纬度植被却吸收了大量的能量。这两个过程齐头并进，意味着热带地区炽热难当，两极则天寒地冻。大自然痛恨这种能量失衡，所以，能量以热能的形式由大气和海洋输送到两极，从而影响了气候。

太空中的地球

气候由地球与太阳之间关系的两个基本事实所控制。第一个事实是地球自转轴的倾斜，这导致了四季的变换。第二个事实是地球每日的自转，这给我们带来了日与夜，并驱动了大气和海洋的循环。

地球自转轴的倾斜角度是23.5°，这导致两个半球全年接收的能量存在季节性差异。目前，季节性变化对气候的影响最大。如果地球不倾斜，而是直直地立在轴线上，那么我们就不会有春夏秋冬了，想到这些不禁令人惊叹。如果真是这样，温带地区的植被不会存在巨大的变化，也不会有热带地区的季风和飓风季节了。四季的由来源于一年中阳光照射地球的角度发生了变化。以12月21日为例，由于地球的自转轴偏离太阳，所以照射到北半球的阳光角度更大，能量扩散的范围也更广。此外，这种倾斜实在是太大了，以至于在北极地区，阳光甚至无法到达地表，因而产生了24小时的黑暗和北半球的冬天。然而，在南半球一切都恰恰相反，因为南半球当时是向着太阳倾斜的，因此烈日炎炎。这意味着南极洲24小时都沐浴在阳光下，澳大利亚人则在海滩上享受圣诞晚餐，同时还能把皮肤晒得更黑。地球绕着太阳运转一圈大约需要365.25天（因此每

6

四年便有一个闰年），轴线的角度保持不变。到了6月，地球的自转轴就会向太阳倾斜，北半球于是有大量的阳光直射，如此便进入夏季，而南半球因远离阳光而坠入冬季。

如果全年追踪太阳，就可以看到这种倾斜是如何通过四季轮回影响地球的。我们从6月21日开始，此时的太阳在正午时分位于北回归线（北纬23.4°）的正上方，即北半球的夏至。此后太阳的角度向南移动，直到9月21日正午时分位于赤道上空，也就是二分点或北半球的秋分。太阳继续南移，12月21日的正午出现在南回归线（南纬23.4°）的正上方，此时是南半球的夏至。然后，太阳看来像是北移了，在3月21日二分点或北半球春分时的正午到了赤道正上方，如此循环下去（图4）。

换季标志着气候中最明显的变化；以纽约为例，冬季的气温可低至−20 ℃，而夏季的气温可以超过35 ℃——温差达55 ℃。此外我们还将发现，季节更替是风暴的主要成因之一。

热量在地球上的移动

影响地球气候的第二个事实是地球日复一日的自转。首先，这使得地球坠入和冲出黑暗，造成昼夜温度的巨大变化。例如，撒哈拉沙漠夏季白天的温度可以超过38 ℃（100 ℉），夜间的最低温度为5 ℃（40 ℉）；而香港的昼夜温差只有4 ℃（7 ℉）多一点。根据季节的不同，不同地区的日照时间也不同。具体的日照时间可能是两极的24小时白昼或24小时黑暗，也可能是赤道每天12小时左右的日照。日照的变化加剧了季节性反差，因为在夏季，我们不仅能获得更多"烈日当头"的阳光，照射的时间也更长了。

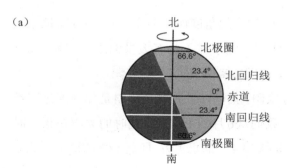

（a）

北

北极圈
66.6°
23.4°　北回归线
0°　赤道
23.4°　南回归线
66.6°
南极圈

南

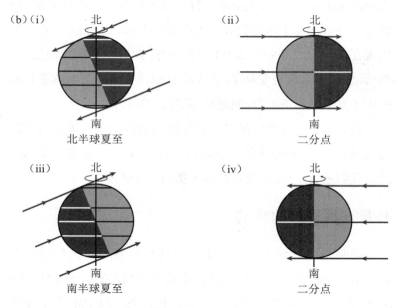

（b）（i）　北

南
北半球夏至

（ii）　北

南
二分点

（iii）　北

南
南半球夏至

（iv）　北

南
二分点

图 4　地球的倾斜导致的二分二至

8　　　　但地球的自转也使从赤道向外输送热量变得更加复杂。这
是因为地球的自转让包括大气层和海洋在内的其他一切都在转
动之中。简单的规律是，地球的自转会使北半球的空气和洋流
被推到其行进方向的右边，而在南半球则被推到行进方向的左
边。这种偏转名为科里奥利效应，越靠近两极，其作用越明显。

日常生活中经常被人引用的一个例子，就是水从排水孔或马桶流下去的方向。在北半球，据说水是按顺时针流下排水孔的，而在南半球则是逆时针。但我不得不告诉诸位，水从浴缸或马桶里排出的方向与科里奥利效应或地球的自转无关，而且目前也没有观察到南北半球的马桶在水流方向上有持续存在的差异。这是因为与水的残留移动和容器形状的影响相比，科里奥利效应的影响小到可以忽略不计。也就是说，赤道地区的奇妙手工作坊向游客展示的科里奥利效应只是一个小把戏而已。例如在肯尼亚穿越赤道时，你会看到竖立的大牌子上写着这一点；如果游客愿意在路边停下来，当地人会兴高采烈地把水从桶里倒进一个大漏斗，似乎能清楚地证明，站在牌子的两侧所看到的水流方向并不相同。不过，这种变化完全取决于演示者的手腕，水怎么倒进去，会影响到它往哪边转。尽管这完全是假的，但我还是喜欢这些演示，因为这意味着许多当地人和游客可以得知有关科里奥利效应的知识！

说回气候，洋流和气流为什么会有这种偏转？试想一下从赤道直接向北发射一枚导弹。因为导弹是从向东自转的地球上发射的，所以导弹也在向东移动。地球自转时，赤道必须在太空中快速移动以跟上其他的地方，因为它是地球最宽的部分。地表向北或南远离赤道时会向内弯曲，所以它不必移动得那么快来跟上赤道。因此在一天之内，赤道必须以每小时1 670千米的速度移动40 074千米（地球的直径），而北回归线（北纬23.4°）必须移动36 750千米，速度为每小时1 530千米，北极圈（北纬66.6°）必须移动17 662千米，所以速度为每小时736千米。北极点根本没有相对运动，所以那里的速度是每小时0千米。一个直观的证明是，如果你和朋友手拉手站在一处同时甩手，手的速度会比你们快很多。因此，从赤道发射的导弹，其东进的速度较快，等同于赤道的东进速度；当它向北移动到北回归线时，地表东进的速度不如导弹快。因为导弹的东进速度快于它所进入的区域，这就给人一种导弹向东北方向移动的感觉。当然越接近两极，这种速度的差异就越大，所以向10 东的偏转就越大（图5）。

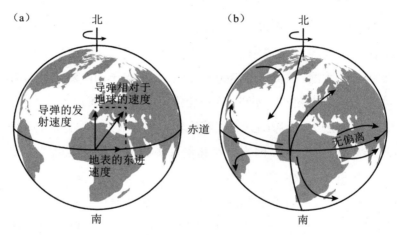

图5　地表的相对运动所产生的科里奥利效应

总　结

气候系统非常简单，它由赤道和两极接收到的不同量的太阳能所控制，气候只是要消除这种不平衡的能量再分配。正如我们将在第二章中看到的那样，进行这种重新分配的正是大气和海洋。由于地球的自转轴与太阳成一定的角度，导致了明显的季节周期，使得情况变得更加复杂。除此之外，地球每24小时就会自转一圈，让地球坠入和脱离黑暗，这也意味着远离赤道的能量的重新分布发生在一个自转的球体上。科里奥利效应由此产生，并有助于解释为什么几乎所有的天气系统似乎都在旋转。 11

大气层与海洋

本章研究大气层和海洋对气候的影响，以及它们如何在全球范围内储存并重新分配太阳的热量。本章将解释海洋为何在热量离开赤道的流动中占主导地位，而中高纬度地区的热流动为什么由大气层主宰。最后将总结世界的主要气候区，并解释为什么全球有三个主要的降雨带和两个主要的沙漠带。

大气层

大气层是天气的发源地。它始于地表，随着高度的上升变得越来越稀薄，与外太空之间没有明确的界限。在62英里（100千米）处随意确定的卡门线通常被用来标示大气层和外太空的边界，它是以匈牙利裔美国工程师和物理学家西奥多·冯·卡门（1881—1963）的名字来命名的。产生天气的大气层在大约10英里的高处开始变得稀薄。海洋在控制天气和气候方面也起着重要的作用。海洋的平均深度约为2.5英里，因此控制气候的物质层的总厚度为12.5英里。

12

大气是气体的物理混合物，而不是化合物。重要的是，这些气体在地表之上约50英里（80千米）处以非常稳定的比例混合。氮气、氧气、氩气和二氧化碳这四种气体占空气体积的99.98%。特别值得关注的是主要由二氧化碳、甲烷和水蒸气组成的温室气体，尽管它们相对稀少，但对大气层的热力特性有很大影响。温室效应将在本章后文中讨论，而全球变暖将在第八章中详细考察。

大气层的成分

氮气是一种无色、无臭、无味、通常不活泼的气体，大约占地球大气层体积的78%。氩气也是一种无色、无臭、无味、完全惰性的气体，大约占地球大气层体积的0.9%。相比之下，氧气是一种反应性很强的气体，大约占地球大气层体积的21%。氧气维持着地球上所有的生命，并在大气层和动植物的生物过程之间不断循环。氧气与氢气结合产生水，水在气态下是水蒸气，就天气而言，它是大气层中最重要的成分之一。

氧气还会形成另一种气体，称为臭氧或三氧，它由三个氧原子组成，而不是通常的两个。这是大气层中极为重要的气体，因为它在平流层（距地表6～31英里之间）形成了一个薄层，可以过滤掉致癌的有害紫外线辐射。然而，即使在这个"层"中，臭氧的含量也只占体积的百万分之二到百万分之八，所以大部分的氧气仍然是正常的二氧类型。这种重要的气体大部分被我们使用的氯氟烃类化合物破坏了，导致南北两极上空出现了臭氧空洞，后来世界各国政府答成共识（例如1985年的《保护臭氧层维也纳公约》和1987年的《关于消耗臭氧层物质的蒙特利尔议定书》），决定停止使用所有氯氟烃类和相关化合物。

二氧化碳占地球大气层的0.04%，是主要的温室气体，对保持地球的相对温暖非常重要。直到近代，二氧化碳的含量一直是通过植物光合作用的消耗和动植物呼吸作用的生成来实现平衡的。不过，近100年来的人类工业发展使得更多的二氧化碳进入大气层，破坏了这种自然的平衡。

气溶胶是海盐、灰尘（特别是来自沙漠地区的灰尘）、有机物和烟雾的悬空颗粒。这些气溶胶所达到的高度将决定它们会造成区域变暖还是变冷。这是因为在大气层的高处，气溶胶有助于反射太阳光，从而使局部地区降温；而在低海拔地区，气溶胶会吸收一些来自地球的热量，因此使局部空气变暖。工业生产增加了大气中气溶胶的含量，导致城市地区出现烟雾、酸雨和局部降温，造成"全球变暗"。但气溶胶最重要的作用是帮助云的形成。没有这些微小的颗粒，水蒸气就不能凝结成云；没有云降水，天气就无从谈起了。

尽管会被人遗忘，但水蒸气是最重要的温室气体，约占大气层体积的1%。不过，水蒸气在时间和空间上变化很大，因为它与复杂的全球水文循环有关。水蒸气在大气中最重要的作用是形成云和产生降水（雨或雪）。暖空气能比冷空气容纳更多的水蒸气。因此，每当一团空气冷却下来，例如空气上升或遇到冷空气团时，就不能容纳那么多的水蒸气，所以水蒸气就会冷凝成气溶胶，形成云。我们稍后要讨论的一个重要问题是，水从气态变为液态时会释放出一些能量，正是这种能量加速了飓风之类大风暴的形成。云的形状、大小各异，是判断即将到来的天气的极佳方式！

14

温室效应

地球的温度是由来自太阳的能量和反射回太空的能量损耗之间的平衡决定的。在地球接收到的太阳短波辐射（主要是紫外辐射和可见"光"）中，几乎所有的辐射都能不受干扰地穿过大气层（图6）。唯一的例外是臭氧，好在它吸收了高能紫外线波段的能量（对我们的细胞有很大的损害），限制了到达地表的能量。大约三分之一的太阳能被直接反射回太空。剩余的能量被陆地和海洋吸收，使它们变暖。然后，它们把获得的这种温暖以长波红外线或"热"辐射的形式发散出去。大气中的水蒸气、二氧化碳、甲烷和一氧化二氮等气体被称为温室气体，因为它们可以吸收部分长波辐射，从而使大气变暖。人们已在大气层中测量到了这种效应，并且可以在实验室中一再重现它。我们需要这种温室效应，因为如果没有它，地球的温度至少会下降35 ℃，届时热带地区的平均温度约为−5 ℃。自工业革命以来，我们一直在燃烧数亿年前沉积的化石燃料（石油、煤炭、天然气），将碳以二氧化碳和甲烷的形式释放回大气中，增加了"温室效应"，并升高了地球的温度。实际上，我们是在把古代储存的阳光重新释放回气候系统中，从而使地球变暖了。

哈德来环流、费雷尔环流和极地环流

如前文所述，地球的形状造成了赤道和两极之间的温度失衡。大气层和海洋都充当了将热量从赤道带走的输送工具。然而，与气候有关的一切都会更复杂一些。在赤道，太阳带来的高温使地表附近的空气变暖，导致空气升入高空，进入

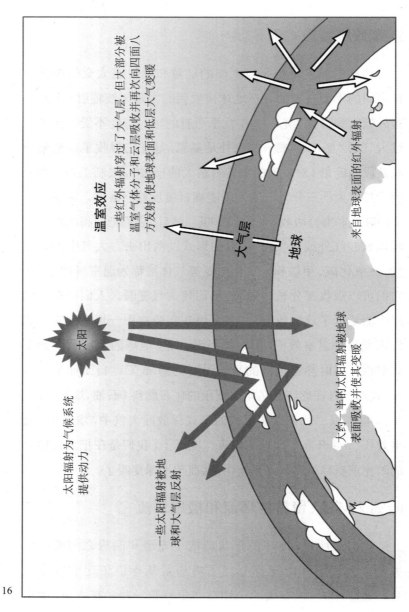

温室效应

太阳辐射为气候系统
提供动力

一些太阳辐射被地
球和大气层反射

一些红外辐射穿过了大气层,但大部分被
温室气体分子和云层吸收并再次向四面八
方发射,使地球表面和低层大气变暖

大气层

地球

来自地球表面的红外辐射

大约一半的太阳辐射被地球
表面吸收并使其变暖

太阳

图 6　温室效应

16

方框1　大气层的垂直结构

大气层通常可以划分为一些界限清晰的层次，温度是主要的划分依据（图7）。

对流层

大气层的最底层是大气湍流和天气现象最明显的区域。它包含大气总分子质量的75%和几乎所有的水蒸气。整层的温度会随着高度普遍下降，平均降幅为每千米6.5 ℃。整个区域的上方是一个逆温层，该层被称为"对流

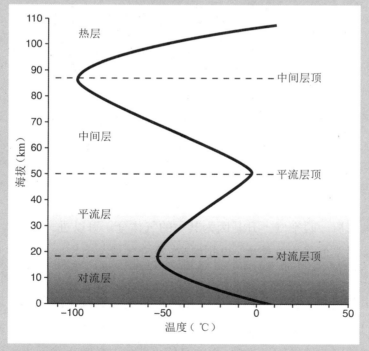

图7　大气层温度分布曲线

层顶"，像是给对流层和天气加上了盖子。

平流层

大气层的第二个主要部分，高度从对流层顶向上延伸至约50千米处。虽然平流层包含了大部分的臭氧，但吸收紫外辐射所致的最高温度出现在"平流层顶"，那里的温度可能超过0℃。温度的大幅上升是由于这个高度的空气密度相对较低。

中间层

在平流层顶之上，平均温度下降到最低-90℃。到了80千米以上，由于臭氧和氧分子对辐射的吸收，温度又开始上升。出现这种逆温的部分被称为"中间层顶"。中间层的大气压极低，从50千米处的100帕降至90千米处的1帕（地表大气压约为100千帕）。

热层

在中间层顶之上，大气密度非常低。由于氧分子和氧原子对太阳辐射的吸收，整个区域的温度会上升。

大气层。暖空气上升是因为其中的气体分子可以进一步分离，从而降低了空气密度，与此相应，冷空气就会下沉。这种向上流失的空气形成了一个空间和低大气压，被吸入的空气填满，由此产生了南北半球的信风。东北和东南信风在热带辐合带（ITCZ）相遇。这造成了一个问题，因为气候系统正竭力从赤道周围的地区输出热量，而吹来的这些风对散热无济于事。因此，在热带地区，大部分热量是由表层流输送的（图8）。墨西

18

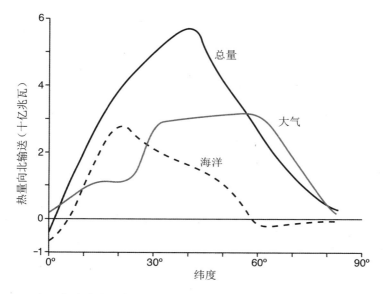

图8　离开赤道的热量输送

哥湾流就是其中之一，该洋流从热带大西洋吸收热量并向北输送，使欧洲的气候全年保持温和（方框2）。其他的洋流包括北太平洋西部的黑潮、南大西洋西部的巴西洋流，以及南太平洋西部的东澳大利亚洋流。

　　然而，在热带地区升入高空大气层的热空气因为上升和向两极移动而慢慢冷却，在南北纬30°左右的地方下沉，形成了副热带高压区。这股下沉气流到达地面时，就会扩散开来向南北两侧移动。该气流失去了大部分水分，使其沉降到达的土地变得干燥，孕育了世界上一些主要的沙漠。南下的空气连接到被称作哈德来环流的第一个大气环流，成为信风系统的一部分。而向北的空气形成了西风带，大气层自此向北取代了海洋，成为热量的主要输送者。北上的亚热带暖空气只有在极锋遭遇极地

19

冷空气团时，才会停止前进。极地的严寒使空气变得极度冷冽而下沉，引起了外吹风（图9）。这股极地冷空气在极锋与温暖湿润的西风带相遇时，交锋导致西风以雨的形式失去大量水分。因为寒冷的极地空气要重得多，它还迫使温暖的亚热带空气上升。这股上升的空气组成了另外两个环流，即费雷尔或中纬度环流，以及极地环流——因为空气上升时会向南北两边扩散。南下的高空空气遭遇北上的热带空气并下沉，形成中费雷尔环流。北上的上升气流飘过极地，在那里冷却下沉，形成极地外吹风，组成了第三个极地环流（图9）。三个环流中的两个由英国律师和业余气象学家乔治·哈德来命名，他在18世纪初解释了维持信风的机制。19世纪中叶，美国气象学家威廉·费雷尔详细解释了中纬度大气环流，发展了哈德来的理论。

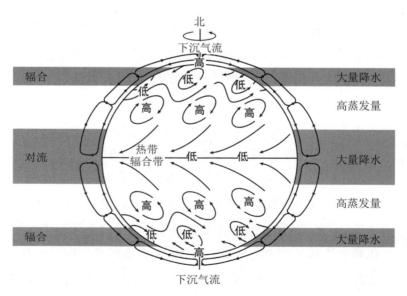

图9 主要的大气环流模式

这些环流的一个重要组成部分是快速流动的高空狭窄气流，称为急流。主要的急流位于代表着对流层和平流层之间过渡的对流层顶附近（方框1）。主要的急流是由西向东流动的西风。它们的路径通常呈蜿蜒状；急流可以开始、停止、分裂成两个或更多的部分，合并成一个气流，或向不同的方向流动，包括与急流的大部队相反的方向。最强的急流是极地急流，在海拔7～12千米左右，海拔更高但稍弱的副热带急流在10～16千米左右（图10）。南北半球各有一股极地急流和一股副热带急流。北半球的极地急流流经北美洲、欧洲、亚洲的中高纬度地区及其间的海洋，而南半球的极地急流常年环绕南极洲。急流是由地球自转和大气中的能量共同作用造成的，因此形成于温差较大的气团边界附近（图10）。

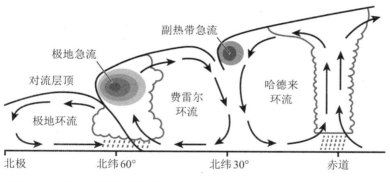

图10　主要的大气环流和急流

虽然地球的一般风型遵循这种简单的三圈环流、每个半球两股急流的模式，但实际上要复杂得多。首先是因为地球在自转，加强了科里奥利效应的影响。这意味着试图向北或向南流动的气团会因地球的自转而偏转。例如，这导致了急流中的大曲流，即所谓的行星波。这些会对天气产生巨大的影响，比如在2012年的春季和夏季，极地急流内的行星波成为静止波，给美国带来了巨大的热浪，也给英国带来了有记录以来最潮湿的4月、5月和6月。其次，大陆的升温速度比海洋快得多，导致陆地上空的地表空气上升，改变了地表风的大气环流。这可以引发局部的海陆风，还在更大的规模上产生了季风系统。因此，季节对大气环流有巨大的影响，因为在两个半球各自的夏季，陆地的升温比海洋快得多，所以，在南半球的夏季，热带辐合带向南被拉向澳大拉西亚①，并穿过南美洲和非洲东南部，在北半球的夏季则向北被拉向印度、东南亚和北非。

然而，哈德来环流确实解释了为什么地球上有三个主要降雨带，即在赤道南北移动的对流降雨带，以及分别在南北半球的两个辐合降雨带，在那里，温暖湿润的亚热带空气与寒冷干燥的极地空气相遇。它们还解释了为什么世界上有两个主要的沙漠带，通常位于降雨带之间，因为哈德来环流和费雷尔环流之间超级干燥的空气会在那里沉降。北半球的典型例子是北非的撒哈拉沙漠和中国的戈壁沙漠，而在南半球，澳大利亚中部的沙漠和南非的卡拉哈里沙漠也与此类似。

哈德来环流也可以用来定义三个主要的风暴区。首先是极锋的"冬季风暴"。其次是副热带高空和信风带，这里是飓风的孕育地。最后是热带辐合带，迅速上升的空气在这里冷却并产生了带来滂沱大雨的热带雷暴，当它移动到陆地上空时，便形成

22

---

① 澳大拉西亚（Australasia），大洋洲的一个地区，包括澳大利亚、新西兰及其附近岛屿。——译注

了季风（详见第四章）。

洋面环流

正如我们所看到的，洋面在全球热量输送中发挥着重要作用。海洋的环流从风开始，因为正是风对洋面的作用才使其移动（图11）。风吹在表层水面上时，摩擦力使能量从风中转移到表层水面上，从而形成了主要的洋流。风能在水柱①湍流中传递到更深的地方，这使得风驱动的洋流可以很深。洋流主要有三种类型：（一）埃克曼运动或输送；（二）惯性流；（三）地转流。

埃克曼运动或输送

沃恩·瓦尔弗里德·埃克曼（1874年5月3日—1954年3月9日）是瑞典的一位海洋学家，他计算得出，在密度相同、无限深广的海洋上有恒定风的情况下，科里奥利效应将是唯一作用在水柱上的其他作用力。离洋面越远，风的影响越小，科里奥利效应就越大，从而导致水的螺旋运动（图12）。其结果是洋面的净运动方向与风向呈90°。弗里乔夫·南森在1890年代的北极探险中最早注意到这一现象，当时他记录到冰的输送似乎与风向成一定的角度。当然，输送的方向取决于所在的半球。在北半球，这种输送的方向在风向的右侧呈90°，而在南半球，则是在风向的左侧呈90°。

---

① 水柱（water column），指从海洋、河流或湖泊表面到底部沉积物的概念性水柱。——译注

24

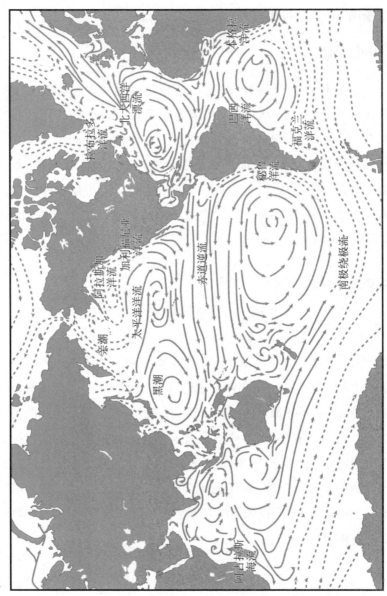

图 11 主要的表层流

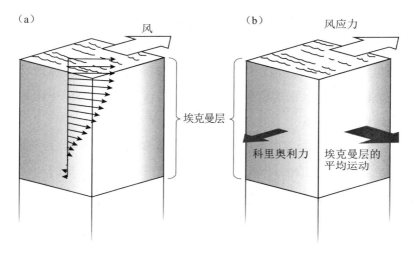

（a） 风

（b） 风应力

埃克曼层

科里奥利力　埃克曼层的平均运动

图12　风力作用下的埃克曼洋面运动

惯性流

　　表层水团体积庞大。例如,墨西哥湾流的水量约为100斯韦德鲁普（100 Sv）。1斯韦德鲁普水量等于每秒10^6立方米或每秒100万吨水。整个世界从河流入海的全部淡水输入量约等于1斯韦德鲁普。这些水团具有巨大的动能,而这就是为什么当风停止推动后,水流仍会继续流淌很远的原因。风停之后,只有摩擦力和科里奥利效应继续作用于水团。如果水团的纬度不变,那么洋流将沿着纬度线流动。如果它的纬度发生了变化,那么科里奥利效应就会起作用,洋流的路径就会急剧弯曲。

地转流

　　与埃克曼的假设相反,海洋并非无限深广。海洋有边界,那就是大陆,而被风吹动的水往往会"堆积"在海洋的一侧,与大

25

陆相抵。这就造成了海面的倾斜，影响了静水压力，导致水从高压区流向低压区。这种力被称为水平气压梯度力，同时也受到科里奥利效应的影响，产生了所谓的地转流。研究地转流的一种方法是观察海面的位势高度——换句话说，就是海洋中比其他地区高的区域。

风力驱动之下的埃克曼流、惯性流和地转流的结合产生了全世界海洋的大部分主要环流特征（图11）。其中的一个主要特征是每个洋盆中的洋涡。这些大型的旋转洋流系统存在于南北大西洋、南北太平洋以及印度洋。不过，还有一种对洋面环流的影响，那就是当深层流形成时，表层水下沉所产生的拉力。

深海环流

深海环流是全球气候的主要控制因素之一，因为它能够在两个半球之间交换热量。事实上，由于深海的体积、热容量和惯性，它是唯一有可能推动和维持内部（几百年到数千年）长期气候变化的因素。现在，太阳在热带地区加热了墨西哥湾的表层水。这种热量也导致了大量的蒸发，将水分送入大气层，开始了水循环。所有这些蒸发使得表层水富含盐分。这些又热又咸的表层水被风从加勒比海沿佛罗里达海岸推到了北大西洋，著名的墨西哥湾流由此开始（图13）。墨西哥湾流的规模大约是亚马孙河最宽处的500倍，它沿着美国海岸流过，然后横跨北大西洋，经过爱尔兰海岸和冰岛，北上进入北欧海域。墨西哥湾流向北流动成为北大西洋漂流，随后冷却下来。高含盐量和低温的结合，增加了表层水的密度或重量。

26

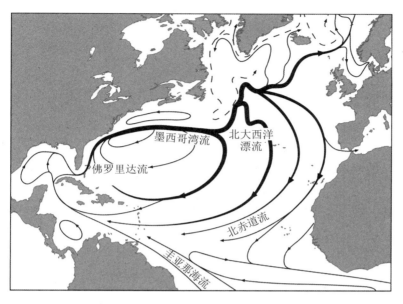

图13 北大西洋主要表层流

现在让我们来看看淡水和海水的区别。淡水冷却后，会发生一些惊人的变化——温度降到4 ℃之前水的密度不断变大，之后变小，然后在0 ℃下结冰。这意味着池塘会从顶部开始结冰，因为密度最大的水位于底部，那里的温度为4 ℃，从而完美地保护了池塘湖泊内的一切生命体。如果一点一点向水中添加盐，它的冰点就会下降，因此我们可以在道路上撒盐防止结冰，但与此同时，密度最大时的温度也会下降（图14）。每千克水中含有26克盐时，其密度最大时的温度与冰点一致。这意味着每千克中含35克盐的海水密度会持续变大直至结冰。水结冰时，会发生另一个惊人的变化——形成了冰这种比其液态物质密度更小的固体。

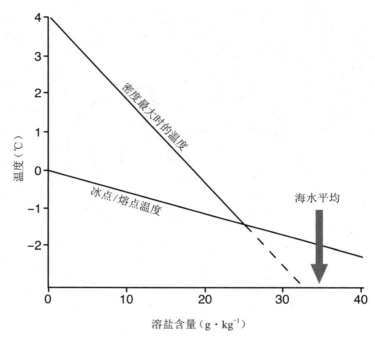

図14 水的温度、盐分和密度的关系

表层水到达冰岛以北盐分相对较低的海洋时已经充分冷却，密度足以沉入深海。这种密集水团下沉所产生的"拉力"有助于维持温暖的墨西哥湾流的强度，确保热带暖流流入大西洋东北部，并将温和的气团送到欧洲大陆。据计算，墨西哥湾流输送的能量相当于一百万座核电站。如果你对于墨西哥湾流给欧洲气候带来的好处有任何疑问，可以比较一下大西洋两岸同一纬度地区的冬季，比如伦敦与拉布拉多，或者里斯本与纽约。或许更明显的是西欧和北美西海岸的差别，它们在海洋和大陆之间有类似的地理关系——例如，阿拉斯加和苏格兰，它们的纬度大致相仿。

北欧海域新形成的深层水在海洋中下沉到 2 000 ～ 3 500 米的深度，并在大西洋里向南流去，称为北大西洋深层水（NADW）。它在南大西洋与第二种深层水相遇，这种深层水形成于南大洋，称为南极底层水（AABW）。南极底层水与北大西洋深层水的形成方式不同。南极洲被海冰包围，深层水形成于海岸冰间湖（海冰上的大洞）。南极的外吹风将海冰推离大陆边缘，产生了这些洞。寒风凛冽，暴露的表层海水被过度冷却。这导致了更多海冰的形成和脱盐，因为结冰的时候会排斥冰水所含的全部盐分，世界上最冷最咸的水由此产生。南极底层水环绕南极流动，并穿透北大西洋，流淌在较暖和因而较轻的北大西洋深层水之下。南极底层水还流入了印度洋和太平洋。北大西洋深层水和南极底层水构成了全球海洋大传送带的关键要素（图15），使两个半球之间能在数十万年的时间里交换热量。

北大西洋深层水和南极底层水之间的平衡对维持当前的气候极为重要，因为它不仅使墨西哥湾流流经欧洲，而且维持了南北半球之间适量的热量交换。根据计算机模拟和对过去气候的研究，科学家们担心，如果输入的淡水量足以使表层水过轻而无法下沉，深层水的环流可能会被削弱或"关闭"。科学家们创造了"去密集化"这个词，意思是通过增加淡水和（或）提升水温来降低密度，这两种方法都能阻止海水因高密度而下沉。人们担心气候变化会造成格陵兰岛部分地区融化，这可能导致更多的淡水流入北欧海域，从而削弱北大西洋深层水和墨西哥湾流。这将给欧洲带来更加寒冷的冬季，天气普遍会变得更加恶劣。不过，由于温暖的墨西哥湾流的影响主要在冬季，这种变化不会影响夏季的气温。所以，就算墨西哥湾流失效，全球变暖仍会导

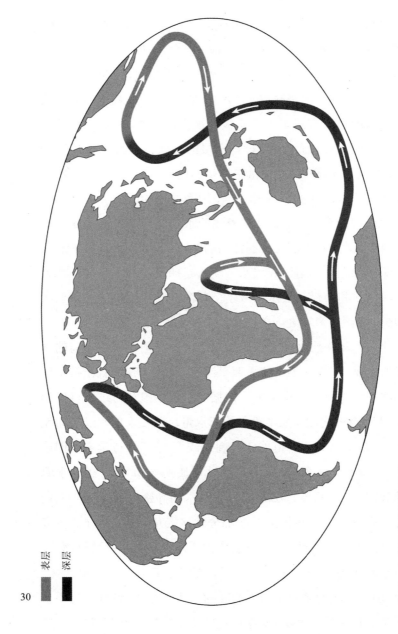

表层

深层

图 15 全球深海传送带环流

致欧洲的夏季变热,欧洲最终会出现与阿拉斯加非常相似的季节性极端天气。

方框2　金发与海洋环流

　　墨西哥湾流还可能给我们带来金发白肤的人。墨西哥湾流对西欧的暖化效应非常显著,这意味着早期的农艺师可以在位于极北之地的挪威和瑞典等国种植农作物。这些早期定居者生活在北至北极圈的地方,该地与格陵兰冰盖中部或阿拉斯加北部冻原处于同一纬度。但生活在如此遥远的北方有一个重大的缺点,那就是缺乏阳光。人类需要维生素D,如果没有维生素D,儿童就会出现佝偻病,引起骨骼软化,导致骨折和严重畸形。维生素D是在皮肤受到紫外线照射时产生的。当然,对于我们演化于非洲的祖先来说这不是问题——情况恰恰相反,深色皮肤是抵御强烈阳光的必要保护。然而,随着我们的祖先向北越走越远,阳光越来越少,产生的维生素D也越来越少,每一代人只有那些皮肤和头发颜色最浅的人才能避免得佝偻病,因为皮肤和头发的颜色越浅,能吸收的阳光就越多,因而能制造的维生素D也就越多。所以在这些地区有很强的选择压力,有利于皮肤白皙、发色金黄的人。另一方面,维生素D也存在于脂肪含量高的鱼类和蘑菇等食物中,这可能是同样的选择压力为何不适用于北极因纽特人的原因。但有意思的是,如果不是墨西哥湾流和斯堪的纳维亚早期移民的固执(他们只依靠农作物,很少吃鱼或根本不吃鱼),人间就没有金发尤物了。

全球植被

世界上的植被区受控于气温和降水的年均值与季节性。温度随纬度而渐变,热带地区最温暖,两极地区最寒冷。正如我们所看到的,地球上有三个主要的降雨带,分别是热带地区的对流降雨带和南北半球中纬度地区的辐合降雨带,世界上两个主要的沙漠区均位于这些降雨带之间。植被随这些气候带的变化而变化。因此热带雨林出现在热带地区,那里一年四季都有大量的降雨。稀树草原位于有高度季节性降雨,但也有持续四个月以上的漫长旱季的热带地区。世上最大的沙漠出现在中纬度地区,在这里,降雨的季节性至关重要,因为虽然许多沙漠的降雨量和伦敦等地一样,但这种降雨只出现在很短的时间段内,一年中的其他时间都极度干旱。当降雨只出现在冬季,随后是非常干燥的夏季时,就会出现独特的地中海植物群落,比如美国加利福尼亚州、南非,当然还有地中海周围。中纬度地区纬度偏高的部分有温带森林或北方森林。年降雨量少的地区有草原植被。温度成为限制因素的高纬度地区有冻原。其他因素也会影响不同植被的存在,例如,我们已经看到,主要的洋流可以使温带气候的植被存在于比通常预期更北的地方。我们将在第五章看到,山脉和高原对降雨发生的地方有巨大的影响,从而影响了沙漠的形成。

最后应该记住的是,植被本身也会影响气候。首先,植被会改变一切地区的反照率,因此热带雨林吸收的太阳辐射比冻原吸收的要多得多。其次,植被非常善于循环利用水分,能够保持大气的潮湿。例如,亚马孙盆地50%的降雨量来自树木循环利用水、蒸发并形成的新云朵。

气候

第三章

天气与气候

引　言

　　许多人把天气和气候混为一谈。要求科学家们预测未来50年的气候时，这种混淆就更加严重了，因为人人都知道，他们连区区几周后的天气都尚且无法预测。所以，气候一般被定义为"平均天气"。气候最初的定义是"30年的平均天气"，这个定义已经变了，因为我们现在知道气候正在发生变化，在过去的50年里，气候每十年都会发生重大的变化。天气的混沌性质使它在几天之后就变得不可预测，而理解气候和模拟气候变化则要容易得多，因为要处理的是长期的平均数。一个很好的对比是，虽然无法预测任何一个人的死亡年龄，但我们可以很有把握地说，发达国家的人平均寿命大约是80岁。另一对被混淆的事情是，人们总是记得极端天气事件，而不是平均天气。比如大家都记得2003年出现在英国和2012年出现在美国的热浪，或者2010年发生在巴基斯坦和澳大利亚的洪水。因此，我们对天气的认识

34 被这些事件所扭曲,而不是出于对平均天气或气候的理解。

混沌理论

美国国家气象局每年花费逾十亿美元,确保国家拥有尽可能准确的天气预报。其他国家也在气象机构投入了同样巨大的资源,因为预测天气是一门大生意,正确预测风暴可以节省数十亿美元,挽救许多生命。如今对三四天后的预报和20年前对两天后的预报一样准确。三天前预测的降雨与1980年代中期的一天前预报一样准确。预报山洪暴发的准确率从60%提高到了86%,此外,这些洪水的潜在受害者可以在将近1小时之前得到预警,而1986年的预警时间仅为8分钟。龙卷风预警的准备时间,也就是居民做出反应的时间,从1986年的5分钟增加到了12分钟以上。严重的区域性雷暴和类似的大暴雨一般在18分钟前就能预见到,而不是20多年前的12分钟。70%的飓风路径可以至少提前24小时预测,飓风的登陆范围可以预测到100英里(160千米)以内。

这些都是伟大的成就,但并不能解释为什么以我们的技术和对气候系统的了解,还是不能提前十天、一个月或一年预测天气。此外,想想看,电视上的天气预报有多少次说今天天气晴朗,结果下雨了。那么,预测天气为什么这样难呢?在1950年代和1960年代,人们认为天气预测受到数据缺乏的限制,如果我们能测量得更加准确,清楚地了解基本过程,就能达到更高的预测水平。但在1961年,麻省理工学院的气象学家爱德华·洛伦茨煮了一杯咖啡,从根本上改变了我们对自然系统的思考方式。

35 1960年,洛伦茨制作出了最早的计算机天气模型之一。1961年的一个冬日，洛伦茨的计算模型产生了一些非常有趣的模式，他想更详细地加以观察。于是他走了一条捷径，从中途开始运行。当然，这是一台最早期的计算机，所以他不得不重新输入所有的起始数字。他没有把它们打到小数点后六位（如0.506 127），而是只打了前三位以节省时间和空间，然后就去煮那杯著名的咖啡了。洛伦茨回来的时候，发现天气模式已经和最初运行的时候有了很大的偏差，两者之间没有可以识别的相似性。看来模型对极小的变化非常敏感，千分之一绝非微不足道，反而对结果产生了巨大的影响。这一原创性工作导致了混沌理论的发展。混沌理论告诉我们，大气温度、压力和湿度的极小变化，都会对大规模的天气模式产生重大、不可预测或混乱的影响。

　　然而，混沌理论并不意味着系统内完全缺乏秩序。恰恰相反，混沌理论告诉我们，可以在一定范围内预测天气会是什么样子：例如我们都知道，龙卷风大多发生在5月的美国，而英国冬季潮湿。但涉及更详细的预测时，一切都行不通了，这是由于所谓的"蝴蝶效应"。这个概念是说，像蝴蝶扇动翅膀那么微小的变化都可以对天气产生巨大的影响，例如改变飓风的强度和方向。从尘卷风、暴风，到只有卫星才能看到的大陆规模的涡流，误差和不确定性成倍增长，并通过湍流特征链向上层层叠加。实际上，我们永远不知道天气的哪些微小变化会结合起来产生这些巨大的影响。洛伦茨在他的天气模式中用了12个方程，而现代的天气计算机使用了50万个。但即使是来自位于英国雷丁的欧洲中期天气预报中心的最佳预报也表明，超过四天的天气 36 预报充其量是推测性的，超过一周的天气预报就毫无价值了，这一切都是因为混沌。

所以混沌理论说，我们可以理解天气，能够预测总体变化，但很难预测暴雨和热浪等个别事件。然而跟气象学相比，气候研究有一个很大的优势，因为它只研究平均数，因此混沌理论不会对它造成影响。此外，在模拟未来的气候变化时，我们现在可以理解的是，地球平均温度的上升将使一些天气现象变得更加频繁和剧烈，例如热浪和暴雨事件，而另一些天气现象将变得不那么频繁和剧烈，例如极寒事件和降雪。

年代际和准周期性气候系统

气候系统包含许多周期和振荡，使我们预测天气的能力受到影响。这些周期和振荡包括年代际周期，如北大西洋涛动（NAO）、大西洋多年代际振荡（AMO）、北极涛动（AO），以及太平洋年代际振荡（PDO）。其中第一个是北大西洋涛动，它最初是由英国物理学家和统计学家吉尔伯特·沃克爵士（1868年6月14日—1958年11月4日）在1920年代描述的。北大西洋涛动是北大西洋的一种气候现象，用来表示冰岛和亚速尔群岛之间海平面的大气压差。冰岛低压和亚速尔群岛高压系统的压差似乎控制着整个北大西洋的西风以及风暴路径的强度与方向。这反过来又控制了欧洲会在何时何地下雨。与恩索（ENSO，见下一节）不同，北大西洋涛动主要受控于大气变化。北大西洋涛动与北极涛动关系密切，虽然两者似乎都是以十年为单位变化的，但看起来并没有周期性。即便如此，也不应将北大西洋涛动与大西洋多年代际振荡混为一谈。

大西洋多年代际振荡是十年尺度的北大西洋海面温度变化。过去130年来，在1885年至1900年、1927年至1947年、1951

年至1961年，以及1998年至今，北大西洋的温度一直比平均温度高，而其他时间则低于平均温度。大西洋多年代际振荡确实影响了北半球大部分地区，特别是北美和欧洲的气温及降雨，例如巴西东北部和非洲萨赫勒地区的降雨，以及北美和欧洲的夏季气候。它还与北美干旱发生频率的变化有关，并有可能影响大西洋强烈飓风的发生频率。还有一些不规则或具有准周期性的周期现象，如印度洋偶极子[①]和恩索。在这些周期中，恩索是迄今为止最著名的，下文将详细讨论。

恩　索

　　全球气候中最重要和最神秘的要素之一，是太平洋地区洋流和气流的方向及强度的周期性变化。这种现象最初被称为厄尔尼诺（西班牙语中的"圣婴"），因为它通常在圣诞节出现，现在更多的时候被称为恩索（厄尔尼诺–南方涛动），这种现象通常每三至七年发生一次，可能持续几个月到一年以上。恩索是三种气候之间的振荡："正常"情况、拉尼娜和"厄尔尼诺"。恩索与世界各地的季风、风暴模式的变化，以及干旱的发生有关。例如，1997年至1998年长期的恩索事件造成了全球气候的剧烈变化，包括东非、印度北部、巴西东北部、澳大利亚、印度尼西亚和美国南部的干旱，还引发了加利福尼亚州、南美洲部分地区、太平洋、斯里兰卡和非洲中东部的暴雨。恩索的状况还与大西洋中飓风的位置和发生有关。例如，人们认为，对"米奇"飓风　38

---

　　① 指印度洋异常的气候振荡现象。该现象出现时会使印度洋东西侧的海水温度混乱，改变正常风向。整个气候模式的变化造成一些地区干旱，而另一些地区降雨量又过多，情况类似厄尔尼诺现象。此现象分别有正负两个极端，当印度洋西侧海面温度比正常高时称为正偶极，相反则称为负偶极。——译注

（见第四章）登陆地点的预测不准确，是因为没有考虑到恩索的情况，强烈的信风推动风暴向南拖过美国中部，而不是像预测的那样向西行进。

厄尔尼诺事件是指西太平洋的表层暖水向东移动到太平洋中心（图16）。因此，强对流单体或上升的暖气柱更接近南美洲。这样一来，信风大大减弱，穿过太平洋的洋流也减弱了。这

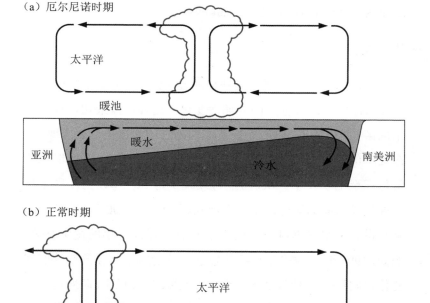

（a）厄尔尼诺时期

太平洋

暖池

亚洲　暖水　冷水　南美洲

（b）正常时期

太平洋

暖池　冷舌

亚洲　暖水　冷水　南美洲

图16　太平洋在厄尔尼诺时期和正常时期的情况

就导致南美洲沿海寒冷且营养丰富的上涌水流量减少，而没有了这些营养物质，海洋中的生命数量就会减少，鱼获量也会急剧降低。这种洋流和上升暖空气位置的巨大变化改变了急流的方向，扰乱了北美洲、非洲和世界其他地区的天气。然而如果想知道是什么原因导致了厄尔尼诺现象，那么答案类似于先有鸡还是先有蛋。是横贯太平洋的西向洋流强度减弱，让暖池①向东扩散，风系随之移动，还是风系强度降低，洋流减弱，从而让暖池东移？许多科学家认为，随着时间的推移，在南美洲和澳大利亚之间移动的太平洋长周期波浪有助于改变洋流，结果要么产生了厄尔尼诺时期，要么会带来拉尼娜时期。

拉尼娜现象是更极端版本的"正常"情况。在正常情况下，太平洋暖池在西太平洋，强烈的西风和洋流使其保持在那里。这带来了从南美洲海岸流出的上涌水流，它提供了大量的营养物质，从而创造了良好的捕鱼条件。在拉尼娜时期，东西太平洋之间的温差达到极限，西风和洋流得到增强。拉尼娜对世界天气的影响比厄尔尼诺更难预测，因为在厄尔尼诺时期，太平洋的急流和风暴轨迹变得更醒目、更笔直，因此更容易预测其影响。拉尼娜现象则弱化了急流和风暴的路径，使其更加迂回而不规则，这意味着大气的活动，特别是风暴的活动，将更加难以预测。一般来说，发生厄尔尼诺现象的地方温暖，出现拉尼娜现象的地方凉爽；发生厄尔尼诺现象的地方潮湿，出现拉尼娜现象的地方干燥。拉尼娜现象曾发生在1904年、1908年、1910年、1916年、1924年、1928年、1938年、1950年、1955年、1964年、1970年、

---

① 热带暖池或称印度洋-太平洋暖池是位于西太平洋和东印度洋的海洋水体，是地球表面水温最高之处。——译注

1973年、1975年、1988年、1995年、1999年、2008年和2011年，其中2010年至2011年的拉尼娜现象是有史以来观测到的最强一次。

预测恩索

厄尔尼诺现象很难预测，但为了更好地了解气候系统，人类已经在过去30年中做了许多工作。例如，现在太平洋上空有一个大型的海洋和卫星监测系统网络，主要目的是记录海面温度，这是恩索状态的主要指标。通过在计算机环流模型和统计模型中使用这些气候数据，能够预测厄尔尼诺或拉尼娜事件发生的可能性。我们对恩索现象的认识和预测能力确实还处于起步阶段。

关于恩索是否受到全球变暖的影响，也存在着大量争论。厄尔尼诺现象一般每三至七年出现一次，但在过去20年中，厄尔尼诺现象表现得非常奇怪，四年中有三年重现了这一现象：1991年至1992年、1993年至1994年和1994年至1995年，随后直到1997年至1998年才再次出现，接下来的八年没有重现，最后一次在2006年至2007年到来。利用西太平洋的珊瑚礁重建过去的气候，可以显示150年来海面温度的变化，远远超过了我们的历史记录。海面温度显示了伴随着厄尔尼诺现象的洋流变化，并揭示出厄尔尼诺事件在频率和强度上的两个重大变化。首先是20世纪初从10～15年的周期转变为3～5年的周期。其次是1976年的阈值发生急剧变化，当时出现了明显的转变，致使厄尔尼诺事件更加强烈，其出现的频率则更是频繁。此外，在过去的几十年里，厄尔尼诺事件的次数增加了，而拉尼娜事件的次数却减少了。即使考虑到年代际对恩索的影响,观察到的数据中恩索变化的规模在过去50年里也似乎增大了60%。

　　然而,正如我们所看到的,不用评估恩索是否会在未来100 年内变得更加极端,单是预测六个月后的厄尔尼诺事件就已经很困难了。大多数关于未来恩索的计算机模型都没有定论;有的会增加,有的没有变化。因此,我们并不知道全球变暖会对气候系统的这一部分产生何种影响。恩索不仅对全球气候有直接的影响,而且还影响到飓风和气旋的数量、强度与路径,以及亚洲季风的强度与时间。因此,在模拟全球变暖的潜在影响时,最大的未知数之一是恩索的变化及其对全球气候系统其他部分的连锁影响。

气候建模

　　整个人类社会都是在了解未来天气的基础上运作的。例如,印度的农民清楚明年的季风降雨何时会来,所以他们明白什么时候该种庄稼了。而印尼的农民知道每年有两次季风降雨,所以他们每年可以有两次收成。这是基于他们对过去的了解,因为在人们的记忆中,季风每年总是在差不多的时间到来。但天气预报的意义比这更深远,因为它影响着生活的方方面面。房屋、公路、铁路、机场、办公室、汽车、火车等都是根据当地的气候设计的。因此,预测未来的气候必不可少,因为我们知道全球变暖正在改变规则,这意味着不能靠一个地区过去的天气来判断未来的天气会怎样。所以,我们必须开发预测和模拟未来的新方法,以便规划生活,使人类社会能够继续充分运转。

　　气候模型有一个完整的层次,从相对简单的沙盒模型到极

第三章　天气与气候

41

其复杂的三维大气环流模型（GCMs）都包括在其中。每一种模型都有助于研究并促进我们对全球气候系统的了解。然而，用于预测未来全球气候的是复杂的三维大气环流模型。这些综合气候模型基于用数学方程表示的物理规律，而这些方程是以全球三维网格来求解的。为了得到最真实的模拟，气候系统的所有主要部分都必须用子模型来表示，包括大气、海洋、陆地表面（地形）、冰雪圈和生物圈，以及各自内部和彼此之间的过程。大多数全球气候模型对这些组成部分有一些基本的表述。将海洋和大气两部分结合在一起的模型称为大气-海洋环流模型（AOGCMs）。

在过去的25年里，气候模型有了极大的改进，这是由于我们对气候系统的了解增加了，但也是因为计算机的能力几乎呈指数式增长。从1990年联合国政府间气候变化专门委员会（IPCC）首次提出到2007年的最新模型，其空间分辨率有了很大的提高。当前这一代大气-海洋环流模型的分辨率为每个点代表110千米见方的区域，到2013年年底[①]的IPCC科学报告发布时，分辨率将变得更加精细。最新的模型或某些团体如今所称的"气候模拟器"囊括了对大气化学、云层、气溶胶过程的更佳表示，以及包括陆地植被反馈在内的碳循环。但模型中最大的未知数或错误并非在物理学方面，而是对未来90年全球温室气体排放的估计。这其中有太多变数，如全球经济、全球和区域人口增长、技术发展、能源使用和强度、政治协议，以及个人的生活方式等。

42

气
候

---

① 本书成书于2013年。——编注

IPCC在2007年的报告中采用选定的未来二氧化碳排放场景运行了逾20个完全独立的大气–海洋环流模型，得出了到2100年可能发生的全球平均温度变化。这与IPCC在2001年的报告相比有很大的变化，那份报告只使用了其中的7个模型。气候模型使用最广泛的潜在排放场景表明，到2100年，全球平均地表温度可能上升$1.1 \sim 6.4\,℃$。根据使用6种最有可能的排放场景的最佳估计，2100年的这个范围是$1.8 \sim 4\,℃$。模型实验表明，即使所有的辐射强迫剂①保持在2000年时的水平不变，未来20年内，温度每十年仍会升高$0.1\,℃$。这主要是由于海洋的反应缓慢。有趣的是，排放场景的选择对2030年之前的温度上升影响不大，因此这是一个非常稳健的估计。所有的模型都表明，未来20年的温度上升速度是20世纪的两倍。重要的是，我们当前在全球排放方面所做的选择将对2030年以后的全球变暖产生重大影响。将于2013年年底公布的下一份IPCC报告虽然会使用大大改进后的排放场景，但到21世纪末，变暖的潜在变化将非常相似。令人惊喜和备感欣慰的是，在过去的25年里，气候模型给我们的答案一直是相同的，这意味着我们确实理解了气候系统，并且可以理解我们过去和未来的行动产生的后果。

---

① 辐射强迫剂（radiation forcing agent），通常称为温室气体。——译注

极端气候

引　言

　　人类可以在从北极到撒哈拉的极端气候中生活、生存，甚至繁衍。我们已经居住在除南极洲外的每一个大陆上，可以通过对技术和生活方式的调整来应对每个地区的平均气候。如果超出当地气候的可预测界限，例如热浪、风暴、干旱和（或）洪水时，就会出现问题。这意味着我们在一个地区定义的极端天气，如热浪，在另一个地区可能被认为是相当正常的现象。每个社会都有一个应对范围，即它可以应对的天气范围：在英国被视为热浪的天气，在肯尼亚可能是夏季的正常情况。然而在气候系统中，最不可预测和最危险的要素之一是风暴。我们将在本章研究风暴是如何形成的，为什么会形成，以及它们的影响。飓风、龙卷风、冬季风暴和季风都将在本章讨论。

飓 风

飓风是始于北大西洋、加勒比海、墨西哥湾、墨西哥西海岸或东北太平洋的强气旋热带风暴（图17）。西太平洋地区称之为台风，印度洋和澳大拉西亚地区就叫它热带气旋。不过，它们都是完全相同的风暴类型，在这里我们将其全部称为飓风。飓风发生在北纬30°至南纬30°之间的热带地区，但不会发生在赤道附近，因为那里没有足够的大气变化产生飓风。要想将风暴归为飓风，持续风速必须超过每小时120千米。当然在完全发展的飓风中，风速可以超过每小时200千米。

飓风是一种横冲直撞的热带风暴，是由大量的雷暴所组成的已经高度组织成环形单体的旋转气团，能在咆哮的风带推动下流通。飓风在海洋上发展成形，一旦登陆，往往会丧失威力——因为飓风与温带风暴不同，是由水的凝结所产生的潜热[①]驱动的。太阳在靠近赤道的地方最强烈，使陆地温度升高，后者继而加热了空气。这股热空气上升，因此从南北两侧吸入空气，产生了信风。季节变化，信风的交锋位置也随之变化，这就是所谓的热带辐合带（ITCZ）。要产生飓风，海面下至少60米处的海水温度必须高于26 ℃，空气湿度必须达到75%～80%。一旦风暴开始，这一组合就能提供适量的热能和水蒸气来维持风暴。例如在北半球的夏季，当热带的北大西洋加热到足够的温度，水分开始蒸发时，就会发生这些情况。最初，温暖的海洋会加热其上方的空气，并使其上升。这就产生了一个从周围地区吸进空

第四章 极端气候

---

① 潜热（latent heat），指物质在恒温相变过程中吸收或释放的热量。——译注

气的低压区。由于热气腾腾的海面明显蒸发,这股上升的空气中含有大量的水蒸气。空气上升后冷却,不再能容纳如此多的水蒸气,结果一些水蒸气凝结成水滴,然后形成了云。这种从水蒸气到水滴的转变会释放出被称作"潜热"的能量。这反过来又会导致空气进一步变暖,并使其上升到更高处。这种反馈可以使飓风内的空气上升到海面以上10 000米的高空。这就成了风暴眼,它所产生的螺旋上升的空气形成了巨大的积雨云云柱。这种现象的袖珍版就是水壶冒出的蒸汽。热空气从水壶中升起时,它会遇到较冷的空气并形成蒸汽,这就是一小片云。如果把手放在蒸汽附近,会感觉到它非常热,这是因为当水蒸气从气态变回液态时,会释放出所有的能量。

飓风内部的空气到达最高处时,就会从风暴眼处向外流动,产生一大片卷云。空气冷却后回落到海平面,在那里被吸回风暴的中心。由于科里奥利力的作用,被吸进飓风底部的空气以顺时针方向旋转进入风暴,而在顶部逸出的空气则以逆时针方向旋转。这种模式在南半球正好相反。飓风形成于距离赤道至少345英里或纬度5°的地方,那里的科里奥利效应足够强,可以使风暴产生所需的螺旋。飓风的规模可以从100千米到逾1 500千米不等。它可以在几天内逐渐形成,也可以在6 ~ 12小时内形成,飓风阶段通常会持续2 ~ 3天,大约需要4 ~ 5天时间才会消失。据科学家估计,热带气旋释放热能的速度为每天50 ~ 200艾焦(10^{18}焦),约相当于1拍瓦(10^{15}瓦)。这一能量释放的速度相当于人类世界能源消耗的70倍,全球发电量的200倍,或每20分钟引爆一枚十兆吨级的核弹。飓风以萨菲尔-辛普森等级来衡量,从一级的热带风暴直到最剧烈的五级。

气候

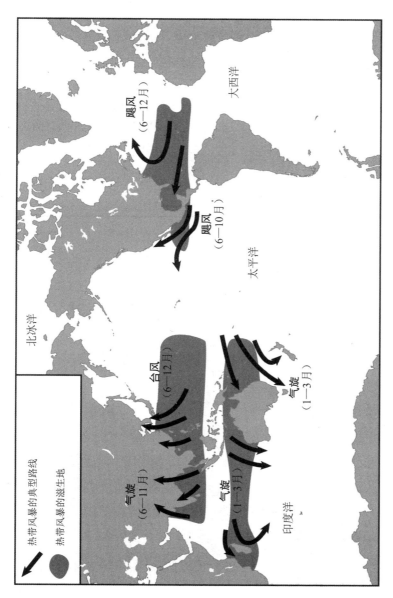

图17 主要热带风暴的位置和发生情况

47

然而，考虑到飓风发生的机会，飓风的形成远比预期的要稀
少得多。热带海洋上空的降压中心中，只有10%能形成完全成
熟的飓风。某个高发年也许最多会有50个热带风暴发展到飓
风级别。灾害的级别难以预测，因为飓风的数量并不重要——
重要的是它们是否登陆（图18）。例如，1992年是北大西洋飓
风非常平静的一年。然而在8月，当年为数不多的飓风之一"安
德鲁"在迈阿密以南袭击了美国，造成的损失估计高达260亿美

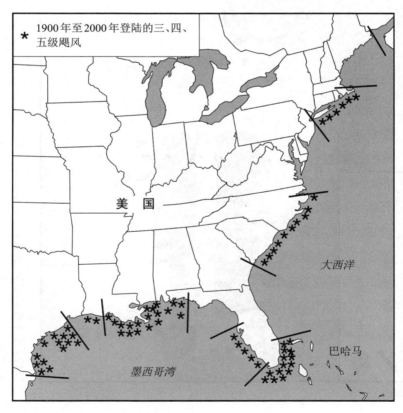

图18　百年以来主要飓风的登陆情况

元。"安德鲁"飓风还表明,预测风暴的袭击地点同样重要——如果飓风袭击了再往北20英里的地方,就会命中迈阿密市人口稠密的地区,造成的损失也会翻倍。

从飓风袭击发达国家的地点来看,主要的影响通常是经济损失,而在发展中国家,主要的影响则是生命损失。例如,2005年袭击新奥尔良的"卡特里娜"飓风造成1 836人死亡,而1998年袭击中美洲的"米奇"飓风至少造成2.5万人死亡,另有200万人无家可归。在这两次飓风中,最大的损失都是由巨量降雨导致的。洪都拉斯、尼加拉瓜、萨尔瓦多和危地马拉受到每小时180英里(290千米)的狂风和每天超过23英寸(60厘米)的降雨量的打击。受灾最严重的洪都拉斯是一个只有600万居民的小国。胡穆亚河通常水面平静,河宽约200英尺(60米),但在飓风的影响下,河水上涨了30英尺(9米),变成了汹涌的洪流,将高度相当于一个城市街区的树木连根拔起。全国85%的地区变成泽国;100多座桥梁、80%的道路和75%的农业设施被毁,包括大部分的香蕉种植园。

在新奥尔良,"卡特里娜"飓风造成的最严重破坏是由强降雨和风暴潮联手造成的。这两方面加在一起,造成53座不同的堤坝决堤,淹没了该市80%的地区。风暴潮还摧毁了密西西比州和亚拉巴马州的海岸。"卡特里娜"飓风并不是美国遭受的最严重的风暴;1926年袭击迈阿密的一场风暴规模要大50%,但由于迈阿密滩当时尚未开发,所以没有造成多少损失。在美国,沿海地区的人口在过去的10~15年里翻了一番,使得这个国家更容易受到风暴相关损失的影响。遭受飓风袭击的发达国家与发展中国家还有很大的经济差异。例如,"卡特里娜"

飓风对美国经济的直接影响逾800亿美元，但它对美国经济的后续影响是，由于布什政府花费了数十亿美元援助该地区的重建，当年还略微促进了美国经济的发展，增幅为1%。而1998年的"米奇"飓风使中美洲的经济倒退了大约十年，两者形成了鲜明的对比。

飓风在世界其他地方也时有发生。每年平均有31个热带风暴在北太平洋西部游荡，从6月到12月，台风会冲向东南亚；风险最大的地区是印度尼西亚、中国和日本，也就是所谓的"台风走廊"。台风走廊为什么会有这么多台风？在这里，它们又为何可以在一年中的几乎任何时候发生？答案就在海洋里。关键是位于西太平洋热带地区的"暖池"海水。信风和洋流将一年四季被太阳加热的热带表层水推到北太平洋的最西边。世界其他地方的飓风季节变幻无定，但"暖池"的海水总是温暖到足以引发飓风——虽然它们在6月至12月之间最为常见（图18）。

龙卷风

龙卷风是大自然中最猛烈的风暴。大气层中没有任何东西要比龙卷风更具破坏性：它们能把一切可以移动的东西卷上云霄；也能把建筑物连根拔起，让疯狂飞舞的残骸形成一团旋涡。它们非常危险，不仅因为风的十足威力及其裹挟的飞射残骸，也因为它们完全不可预测。龙卷风的强度和破坏能力是用藤田级别来衡量的。

龙卷风是一种剧烈旋转的气柱，在远处看起来就像冰淇淋套筒状的云体。在自然界中与龙卷风性质相似的其他风暴还有旋风、尘卷风（龙卷风较弱的同类，发生于陆地）和水龙卷（发生

气
候

在水面的龙卷风）。龙卷风在美国中部、东部和东北部最常见，破坏性也最强，每年5月，平均每天都有5次关于龙卷风的报告。龙卷风在澳大利亚（每年15次）、英国、意大利、日本、孟加拉国、印度东部和中亚也很常见。虽然遇难者人数最多的是美国，但到目前为止，最致命的龙卷风发生在孟加拉国和印度东部这个狭小的范围内。在这个8 000平方英里（21 000平方千米）的地区内已知的42次龙卷风中，有24次死亡人数超过100人。这可能是由于该地区人口密度大，经济状况差，以及缺乏预警系统所致。

我们可以把龙卷风看作微型的飓风。虽然龙卷风也可以在热带海洋上形成，但在陆地上更常见。当地面附近的空气温暖湿润，而其上的空气寒冷干燥时，就有利于龙卷风的形成。这种情况经常发生在美国大平原的春末夏初（图19）。太阳使地面急剧升温，导致温暖潮湿的空气上升。上升的空气在这一过程中冷却并形成大片的积雨云。上升气流的强度决定了周围有多少空气被吸进龙卷风的底部。帮助龙卷风剧烈旋转的因素有两个：其一是科里奥利力，另一个是经过风暴顶部的高空急流，后者给龙卷风添加了额外的螺旋。由于龙卷风形成的条件，它们很容易发生在雷雨和飓风的下方。

尽管美国的某些龙卷风路径曲折，变化迅速，但有近九成是从西南向东北方向移动的。较弱或正在衰减的龙卷风外观呈细绳状。最猛烈的龙卷风外形呈宽大阴暗的漏斗状，从大型雷暴阴暗的云墙中延伸出来。甚至有报道称，有些龙卷风几乎静止不动，始终盘旋在一片田野上，还有的以每小时5英里的速度缓慢前行。而另一些龙卷风的时速超过70英里。但平均而言，

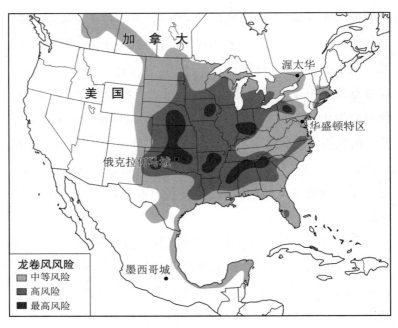

图19　美国龙卷风风险分布图

龙卷风的时速为35英里。人们注意到，大多数龙卷风发生在下午三点到晚上九点之间，但也有在无论白天或晚上的任何时间发生的情况。它们通常只持续15分钟左右，在任何一个地方只停留几秒钟——但还有些龙卷风就是不符合任何规则，例如在1925年3月18日，一股龙卷风在3.5小时内穿越密苏里州、伊利诺伊州和印第安纳州，行程219英里，造成695人死亡。

龙卷风走廊

龙卷风走廊是美国龙卷风发生最多的地区的别称。随着来自太阳的热量不断攀升，天气越来越温暖，源于墨西哥湾的暖湿气流继续向北扩散，龙卷风走廊在春夏两季不断扩大（图19）。龙卷风走廊的核心地区包括得克萨斯州中部、俄克拉何马州和堪萨斯州，但在季节结束之前，它可以向北扩展到内布拉斯加州和艾奥瓦州。龙卷风走廊会随着时间的推移而收缩或膨胀，但始终只有一个。世上没有其他任何地方有组合得如此完美的形成龙卷风的天气条件。形成这个特殊区域的主要原因是：（一）从春天开始持续到夏天，南方和东南方的低层风场把大量的热带暖湿气流从墨西哥湾带到大平原；（二）在离地面大约3 000英尺的地方，有从落基山脉东坡或墨西哥北部的沙漠吹来的非常干燥的气流；（三）在10 000英尺高的地方，盛行西风携带着来自太平洋的冷空气从头顶快速通过，有时还伴随着强大的急流，这提供了巨大的温差，引发了龙卷风和螺旋。

2011年，美国的龙卷风走廊报告了1 897次龙卷风，超过了2004年1 817次的记录。2011年也是龙卷风破坏力特别强、死亡人数特别高的一年，全球至少有577人死亡。其中估计有553人死于美国，而此前十年，美国龙卷风死亡人数的总和为564人。那一年，单年因龙卷风死亡的人数在美国历史上位居第二。然而，这与1989年4月26日发生在孟加拉国的有记录以来最致命的龙卷风相比，仍然相差甚远，那次龙卷风造成逾1 300人死亡，12 000人受伤，从达乌拉特普尔到萨图利亚，除了几棵树之外，其他的一切都被摧毁了。

冬季风暴

对于生活在中纬度地区的人来说，天气似乎是一个永恒的话题。这是因为那里总是在变天。英国有个说法：“如果你不喜欢这个天气,等一个钟头它就会变了。"这是因为中纬度地区的气候是由向南移动的极地冷空气和向北移动的亚热带暖空气之间的大交锋所主导的。这种气团之间的交锋发生在极锋。

54 极锋随着季节的变化而南北移动。夏季亚热带空气比较温暖时,它就会向极地移动。冬季气温较低时,极地气团占主导地位,极锋就会向赤道移动。在这两个气团相遇的地方,就会形成降雨。这是因为暖空气可以容纳更多的水蒸气,它与冷空气交锋时,水蒸气就会凝结成云,从而产生降雨。但真正控制极锋的形状,进而控制极锋天气的是高层大气。高层大气的特点是环绕地球高速移动的快速"急流"。这些强大的急流推动极锋环绕地球,但极锋也随之起了褶皱,形成大量所谓的行星波环绕地球平缓移动。这些波对天气有很大的影响,使我们都在抱怨天气变幻无常,当然还有潮湿。一个波可以在24小时左右经过城市的上空。开始的时候,天气会比较冷,但是天空晴朗。暖锋从头顶经过时,天气会变得更暖和,并开始下雨——通常是小雨或毛毛雨。暖气团的中心抵达城市时,天气转为多云闷热,雨也随之停止。接着,第二条锋线,即冷锋,从头顶掠过;气温下降,并出现短暂的滂沱大雨。随后又恢复到晴冷的天气,直到下个波抵达该城。

气候

正如我们所看到的那样,有许多风暴与哈德来环流一节中描述的不同大气环流区域有关。冰、风、雹和雪暴与极锋或高山地区有关,到了冬季更为严重。在北半球,这些类型的风暴在北美洲、亚欧大陆和日本上空都很常见。

雪要到达地面,云底与地面之间的空气温度必须低于4 ℃,否则雪花穿行于空中时就会融化。要形成雹块,风暴的顶部必

须非常寒冷。高空的水滴处于 0 ℃ 以下的过冷态时,在大气中
碰撞就会形成冰球或雹块。如果切开雹块,可以看到像洋葱一
样堆积起来的冰层。这些雹块从2毫米到20厘米不等,其大小
取决于空气上升气流的强度,因为这决定了它们在大气中停留
多久才会掉下来。最恶劣的风暴天气被称为暴风雪。暴风雪结
合强风,带来了雪、冰和冰雹,气温低至−12 ℃,能见度低于150
米(方框3)。

方框3 身陷严寒

当身体在与寒冷的斗争中败下阵来时,发现这种情况
的往往会是别人。所以,应该时刻注意同伴们暴露于寒冷
天气时的表现。寒冷开始对人造成严重影响的时候,受寒
的人不一定能最好地判断问题的严重性。他仍然认为自己
没事,只是需要再休息一分钟。以下这些都是需要注意的
迹象:

- 寒战无法停止;
- 手部动作笨拙;
- 语速迟缓,口齿不清,甚至可能语无伦次;
- 走路跌跌撞撞,东倒西歪;
- 昏昏欲睡,疲惫不堪,即使在室外也觉得有必要躺下;
- 也许已经休息过了,却还是无法再站起来。

出现这样的行为后,需要穿上干燥的衣服,躺在温暖的
床上。因为此人的体核温度已经开始下降,这对身体来说
非常危险;如果不加制止,将会导致死亡。他们需要温暖

的热水瓶、电褥子，或是给身体盖上热毛巾。他们需要的是热饮，**绝非**含酒精或咖啡因的饮料，因为这些会让人心跳加速，从而丧失更多的热量；这些饮料还会让身体脱水，阻碍恢复。同样**不要**按摩或揉搓此人，因为这也会带走身体核心部位最需要的热量。另外，此人一定要就医。

季 风

大规模暴风雨的另一个重要区域是季风带。季风（monsoon）这个名字来自阿拉伯语mausim，意思是"季节"，因为东南亚的大部分降雨都发生在夏季。在热带地区，因为烈日当头，太阳的能量是最强的。这加热了陆地和海洋，从而使其上方的空气变暖。这些温暖潮湿的空气上升，在其下方留下一个低压区，有助于从周围地区吸入空气（图20）。这种吸力导致了信风，让信风从纬度高得多的地区到达这个空气上升的区域。由于风来自南北两个半球，这个区域被称为热带辐合带。热带辐合带的空气上升时，会形成巨大的塔状云，产生大量的雨水。因为阳光最强烈的位置在赤道上下方移动，热带辐合带也随季节的变化而南北移动。它还受到各大洲位置的强烈影响。这是因为陆地的升温速度比海洋快，幅度也比海洋大，因此在相应的季节，陆地可以将热带辐合带拉向更北或更南的区域。一个例子是亚洲的夏季季风，在夏季，喜马拉雅山脉附近和印度的低地会升温，这就使热带辐合带越过赤道，来到亚洲。因为南半球的风被拉过温暖的印度洋，所以温暖且饱含水分；当它们被迫在印度上空上升并降

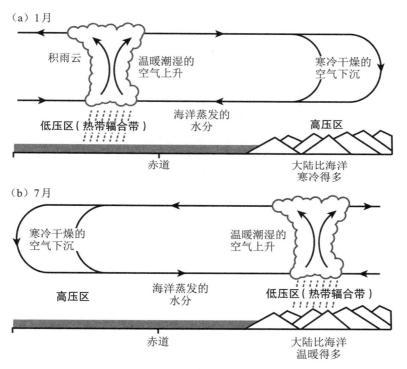

（a）1月

积雨云　温暖潮湿的
空气上升

寒冷干燥的
空气下沉

低压区（热带辐合带）　海洋蒸发的水分　高压区

赤道　大陆比海洋
寒冷得多

（b）7月

寒冷干燥的
空气下沉

温暖潮湿的
空气上升

高压区　海洋蒸发的水分　低压区（热带辐合带）

赤道　大陆比海洋
温暖得多

图20　季风系统

温时，就会在整个东南亚以及北至日本产生非常大的降雨。在北半球的冬季，热带辐合带移动到赤道以南，但在东南亚，这意味着来自北太平洋的温暖、潮湿的风被拖拽着向南穿过大陆，进入南半球。因此，印度尼西亚和中国南方的一些地区每年有两个季风雨季：一个来自北方，一个来自南方。难怪这里是地球上最肥沃的地方，养育了世界上超过五分之二的人口。尽管降雨带来了生机，但也会造成危害，特别是以洪水的形式出现的灾难。这方面的例子是1998年在孟加拉国和中国发生的可怕洪水，造成三百多亿美元的损失和数千人死亡。

57

亚马孙季风

在南半球的夏季,南美洲大陆会升温。这股上升的空气在地面留下了一个低压区,通过吸进周围的空气来填补。这就拉着热带南北部之间的辐合带空气南移,一直拉到巴西的上空。热带辐合带的南移带来了大量的降雨,因为源自温暖的热带大西洋的空气从北方被拉过了赤道。亚马孙季风由此产生,并形成了世界上最大的河流和地球上范围最大的雨林。亚马孙盆地的面积达到惊人的270万平方英里,大部分被雨林所覆盖。在流入全世界海洋的所有淡水中,亚马孙河的水占20%。如果没有季风降水,就不会存在世界上最多样化的栖息地了。

生活在亚洲季风下

孟加拉国是一个名副其实的季风国家,因为四分之三以上的国土是由恒河、布拉马普特拉河和梅克纳河带来的沉积物形成的三角洲地区,所有这些河流都由夏季季风提供水源。国家逾半的区域位于海拔不到5米的地方,因此经常发生洪水。在正常的夏季季风期间,全国有四分之一的地区会被洪水淹没。然而,这些洪水和尼罗河的洪水一样,既造成了破坏,也带来了生机。水能用于灌溉,淤泥能使土地肥沃。肥沃的孟加拉三角洲是世界上人口最密集的地区之一,在14万平方千米的土地上,有超过1.1亿的人口。但每隔一段时间,季风洪水就会超过孟加拉国所能承受的程度。1998年,该国四分之三的地区被洪水淹没了两个月,造成价值数十亿英镑的损失和数千人死亡。孟加拉国还得应对热带气旋。如果我们把热带气旋最严重的三个年份拿出来比较,可以看到伤亡人数急剧下降。1970年,与气旋有关的死亡人数超过30万,1991年为13.8万,而2007年只有3 500人死亡。这绝非热带气旋变温和了,而是因为良好的治理。孟加拉国政府投资建设了优良的气象设施,尽可能准确地预测气旋会在何时何地登陆;此外,他们建立了一个自行车通信网络,一旦发出气旋警报,消息就会传递到所有将受影响的城镇和乡村。他们还修建了气旋避难所,保护了供水和卫生设施,并鼓励浮动化农业①,从而抵御风暴的侵袭。这些相对简单的改变最终拯救了数十万人的生命。

---

① 指孟加拉国适应气候变化和可持续社区发展的潜在清洁生产技术。浮动化农业是一种在常淹没区种植作物的农耕方式,将浮于水面上且充满养分的腐殖质视作土壤,因此这种农耕技术不受洪水泛滥的影响。——译注

构造与气候

引　言

　　在第一章和第二章中，我们看到了气候是如何随着太阳的能量落在地球上，然后在全球范围内重新分配而变化的过程。这两个方面都受到板块构造的强烈影响。这就是为什么一亿年前的地球要温暖湿润得多，恐龙能在南极洲快乐生活的原因。现代的气候系统是数百万年板块构造的产物，板块构造引起了诸多独特的现象，比如两极都有大量的冰。这产生了非常强烈的赤道-极地温度梯度，从而形成了一个极富活力和能量的气候系统。构造对气候有两个主要影响。第一是直接的影响，包括改变了大气环流的山地和高原抬升，以及改变了海洋循环方式的水文循环或海洋通道。第二是间接的影响，通过俯冲、火山活动和消耗气体的化学风化作用，影响大气的成分。贯穿本书的一个主题是气候学并不复杂。构造对气候的影响也是如此。在这一章中，影响因素被分解为水平构造，它研究的是如果大陆板

块在全球范围内移动,会发生什么。其次是垂直构造,它研究如
果创造一座山或一片高原,会发生什么。最后,我们将研究火山
和超级火山对气候的影响。

水平构造

横向大陆

　　大陆的南北位置对两极和赤道之间的热力梯度有巨大的影
响。地质学家运行了简单的气候模型来观察这种影响(图21)。
如果把所有的大陆都放在赤道周围,也就是在所谓的热带"环
形世界"中,两极和赤道之间的温度梯度约为30 ℃(图21)。因
为当两极被海洋覆盖时,其温度无法低于冰点。这是由大气和
海洋二者的特点造成的。气候的一个基本规则是热空气上升,
冷空气下降。两极天气寒冷,所以空气下降,落到地面上就会从
极点向外推。极点的海水结冰后形成海冰,这些冰随即会从极
点被吹向温暖的水域,在那里融化。如此便维持了平衡,使两极
的温度不会低于零度。然而,只要把陆地引入极点,即便只是在
极点周围,都会常年结冰。如果极点上确实有一块像南极洲这
样的陆地,其上有冰,那么赤道–极地的温度梯度就会超过65 ℃
(图21);这正是我们面对的现实情况。相反,如果细想一下北
半球的情况,大陆并不在极点,而是围绕着它。因此,格陵兰岛
的冰盖较小,并没有像南极洲的那样巨大,而各大陆就像一道栅
栏,把所有的海冰都挡在北冰洋上。所以北半球的赤道–极地温
度梯度为大约50 ℃,介于南极洲与无冰大陆这两个极端之间。
赤道–极地温度梯度的大小是气候的基本驱动力。因为海洋和
大气环流的主要驱动力把热量从赤道转移到了两极,所以这个

63

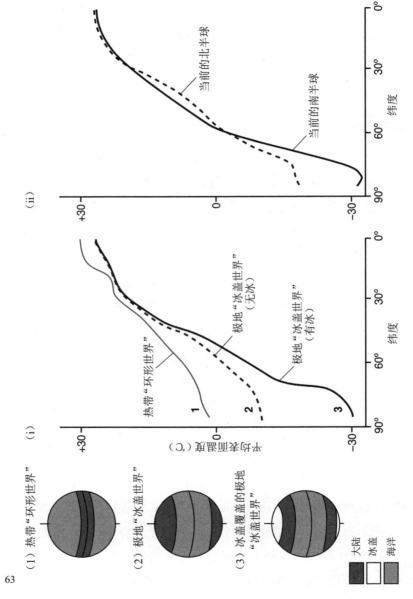

(i)

+30

中等温度 0
温度（℃）

−30

0° 30° 60° 90°
纬度

热带"环形世界"
1

极地"冰盖世界"（无冰）
2

极地"冰盖世界"（有冰）
3

(ii)

+30

0

−30

0° 30° 60° 90°
纬度

当前的北半球

当前的南半球

(1) 热带"环形世界"

(2) 极地"冰盖世界"

(3) 冰盖覆盖的极地"冰盖世界"

大陆
冰盖
海洋

图 21　大陆的纬度位置和赤道−极地温度梯度

温度梯度决定了世界会有怎样的气候。寒冷的地球有极端的赤道-极地温度梯度，气候也就变幻多端。这就是存在强烈的飓风和冬季风暴的原因：气候系统正试图将热量从炎热的热带地区输送到寒冷的两极。

纵向大陆

本书第二章介绍了海洋环流的基本原理。海洋环流的一个关键方面在于海洋是如何被抑制的。如果没有大陆的阻挡，那么海洋就会持续绕着地球循环。然而当洋流遇到大陆时，就会向南北两个方向偏转。如果我们看一下现代大陆的布局［图22（a）］，就会发现有三个主要的纵向大陆：（一）南北美洲；（二）欧洲到非洲南部；（三）东北亚到澳大拉西亚。即使在一亿年前，这些大陆仍可辨认出来，但位置略有不同［图22（b）］。当时的布局有两个显著的特点：第一，有一片经特提斯海和中美洲深海通道穿过整个热带地区的海洋；第二，南极洲周围没有海洋环绕。这些变化对洋面环流，进而对深海环流和全球气候产生了巨大的影响。理解海洋通道对海洋环流的影响主要有三种概念性的方式。第一种是简单的双片世界，两边都有纵向的大陆（图23）。由于洋流在热带和两极受地面或洋面风的驱动被推向西边，而在中纬度地区则被推向东边，这就在两个半球产生了双流涡的经典结果。如今北太平洋和北大西洋都有这种类型的环流。第二种是有低纬度海道的双片世界。这产生了绕着世界不断向西流淌的一片庞大的热带海洋，从而使每个半球都有两个较小的流涡（图23）。这就是白垩纪时期的环流，每个半球的两个流涡都在太平洋上。第

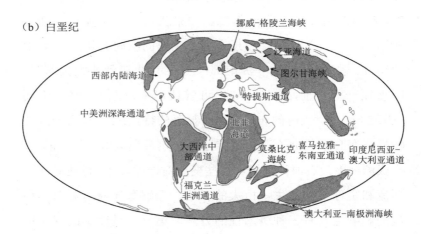

（a）现代

弗拉姆海峡

白令海峡

丹麦海峡

冰岛-法罗群岛通道

直布罗陀
海峡

吕宋海峡

巴拿马地峡

印度尼西亚海峡

塔斯曼-南极通道

德雷克海峡

（b）白垩纪

挪威-格陵兰海峡

泛亚海道

西部内陆海道

图尔甘海峡

特提斯通道

中美洲深海通道

北非
海道

大西洋中
部通道

莫桑比克
海峡

喜马拉雅-
东南亚通道

印度尼西亚-
澳大利亚通道

福克兰-
非洲通道

澳大利亚-南极洲海峡

图22 现代和白垩纪时期的海洋通道

气
候

三种是有高纬度海道的双片世界。这在两个半球都产生了强烈的绕极洋流，每个半球也都有一个单一的热带流涡（图23）。当今的南半球类似于这种情况，南极洲的周围有绕极洋流。因此，南大洋就像一个巨大的海洋吸热器，对南极洲冰层的大量堆积起到了重要作用。

65

64

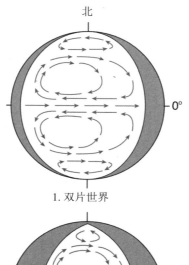

1. 双片世界

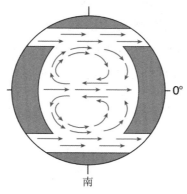

2. 有低纬度海道的双片世界

3. 有高纬度海道的双片世界

图 23　纵向大陆和海洋环流　　　　　　　　66

深海环流

　　深海环流也是一个重要的考虑因素,因为它影响着洋面的环流和两个半球之间的分配。海洋通道的存在与否对深海环流有着深远的影响。例如,如今帮助将墨西哥湾流拉向北方,维持了欧洲温和气候的北大西洋深层水可能只有400万年的历史。

如果我们运行海洋环流的计算机模拟,不管有没有德雷克海峡和巴拿马通道,只有眼下这种组合才会产生显著的北大西洋深层水。因此,当今的深海环流是由大约2 500万年前开放的德雷克海峡和大约400万年后关闭的巴拿马通道造成的(图24)。这都是因为盐的缘故。由于北大西洋地区的蒸发效果更强烈,北大西洋比太平洋的盐度更高。来自加勒比海的温暖咸水穿越大西洋并降温,就形成了今天的北大西洋深层水。高含盐量和低温共同作用,增加了水的密度,使它能够在冰岛以北沉降。所以当巴拿马通道开放时,较淡的太平洋水就会渗入,降低了北大西洋的整体含盐量。这样一来,洋面水即使冷却了,其密度也不足以下沉,所以与当今的情况相比,就不能形成这么多的北大西洋深层水。因此,现代气候系统的基本要素之一,即南极底层水和北大西洋深层水之间的竞争,原来是一个新兴的特征。

垂直构造

　　构造板块在地球表面移动时经常彼此碰撞,每当这种情况发生时,陆地就会被推向高处。这在某些情况下会形成山峦,或者整个地区被抬高形成了高原。这些都对气候系统产生了深远的影响。其中的一个是雨影,即山系背风面的干燥区域。通常

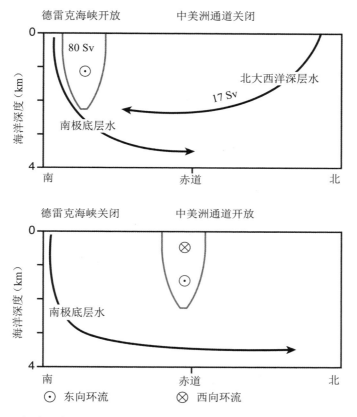

图24 海洋通道与深海环流

在山系前侧有一片相应的降水增加的区域。地面的天气系统向山地或高原移动时,通常相对温暖和潮湿(图25)。遇到山体后,空气被迫向上移动并越过山体。由于大气压力随着海拔高度的增加而降低,空气必然膨胀,同时冷却下来。冷空气比暖空气能容纳的水分要少,所以相对湿度迅速上升,达到100%时便发生了强降雨。空气下降到山的另一侧时,大气压增加,空气温度上升,相对湿度下降得很低,因为空气中只有很少或完全没有

68

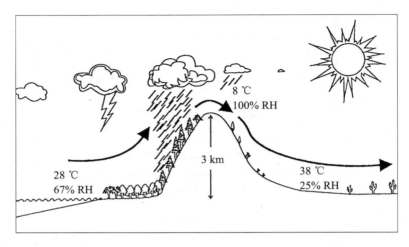

图 25　山脉的雨影（RH＝相对湿度）

水分。因此，在下降的一侧有一片雨影，因为没有留下形成雨的
水分，由此可能导致沙漠的产生。这个简单的过程可以控制整
个大陆的湿润或干燥程度。图26显示了山脉出现在大陆的西
部或东部边界的影响。正如我们在第二章中所看到的，世界上
有三个主要的降雨带，一个在热带，另外两个分别在两个半球的
中纬度地区。热带的空气由东向西移动，而中纬度地区的空气
则由西向东移动。所以大陆西侧的山会在陆地上带来更多的降
水，大陆总体上也会更加湿润。巧合的是，当今的北美洲西海岸
有纵向贯穿的西部山脉，即落基山脉，而南美洲西海岸则有安第
斯山脉。这些山脉不仅孕育了重要的湿润地区，还产生了著名
的沙漠，比如地球上最干燥的两个沙漠——智利的阿塔卡马沙
69　漠和美国的死亡谷。如果抬升形成了高原，湿润地区和干燥地
区的对比就更加鲜明了。图26显示了由于这种雨影效应，极少
的降水能够进入高原。

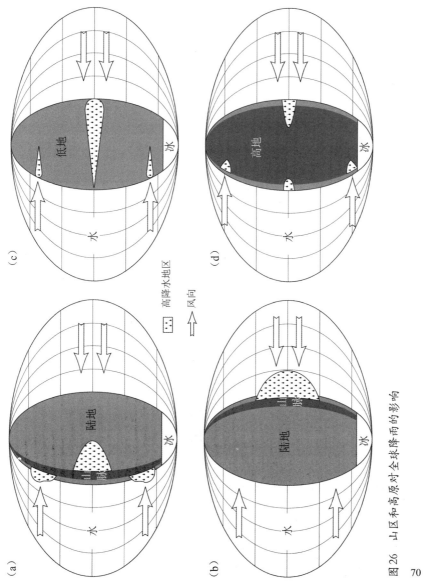

图 26　山区和高原对全球降雨的影响

69

70

大气层屏障

　　巨大的山脉或高原被推得直冲云霄时，就会干扰大气的循环。它们不仅迫使空气上升并越过它们，而且在许多情况下，还会导致周围的气候系统发生偏移。这种影响更加复杂，因为与周围的低地相比，隆起的地区在夏季升温速度更快、幅度更大，在冬季降温速度更慢、幅度更小。图27显示，如果北半球的所有大陆都是平坦的，那么大气的主要环流将接近圆形，同时由于陆地和海洋之间的差异，也许会有轻微的偏转。但是，如果考虑到现今的两座高原，也就是喜马拉雅山-青藏高原和谢拉山-科罗拉多高原的隆起区域，那么环流就会发生巨大的变化。这两座高原都体量庞大：青藏高原是世界上海拔最高、面积最大的高原，达250万平方千米，约为法国面积的4倍；而科罗拉多高原的面积为33.7万平方千米，并与其他众多的高原相连，构成了谢拉山-科罗拉多隆起复合体。

　　在北半球的夏天，这两座主要的高原都比周围地区更热，因此其上的空气上升，出现一个低压区。这就吸进了周围的空气，造成气旋环流，使气候系统更多地向北和向南偏移。在北半球的冬季，这些高地比周围地区冷得多，形成高压系统和外吹反气旋环流（图27）。这使北极空气向北偏移，让亚洲和北美大陆中部地区比那里原本应有的气候情况更温暖。当格陵兰岛、北美洲和欧洲存在大面积冰盖时，大气环流变得更加复杂。因为冰盖总是很冷，所以会产生具有外吹反气旋环流的永久性高压系统，我们将在第七章中讨论这一现象。喜马拉雅山-青藏高原周围的夏季气旋环流也形成了东南季风系统。部分被拉向喜马拉

71

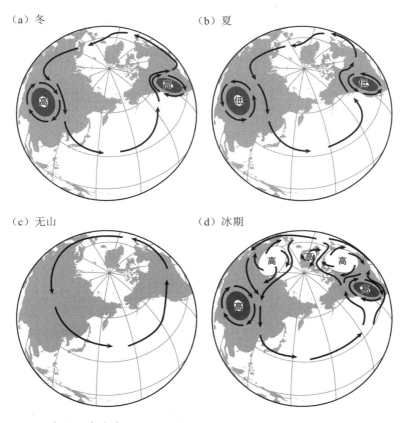

（a）冬　　　　　　　　（b）夏

（c）无山　　　　　　　（d）冰期

图27　高原和冰盖对大气环流的影响

雅山脉的空气来自印度洋，所以带来了大量的水分，由此产生的降雨对世界五分之二人口的福祉至关重要。

火山爆发

　　板块构造控制着火山的发展，火山通过将气体和尘埃引入大气层，对气候产生了重要的影响。一般规模的火山会向对流层喷射二氧化硫（SO_2）、二氧化碳（CO_2）和尘埃，对我们的天气

72

产生相当大的影响。例如，1883年的喀拉喀托火山爆发就造成了36 417人死亡。这次爆发被认为是现代史上人们听到的最响亮的声音，有报道说在近3 000英里之外都能听得到。它相当于2亿吨级的TNT炸药，约为第二次世界大战期间摧毁日本广岛的核弹"小男孩"当量的1.3万倍。注入大气层的二氧化硫和尘埃增加了反射回太空的太阳光，火山爆发后的一年里，全球平均气温下降了1.2 ℃之多。天气模式持续混乱多年，直到1888年气温才恢复正常。

1991年6月15日，皮纳图博火山爆发，向大气层排放了2 000万吨二氧化硫。二氧化硫在大气中氧化，产生了含硫酸（H_2SO_4）液滴的雾气，在爆发后的一年中逐渐扩散到整个平流层的下部。这一次，现代的仪器能够测量其影响，比如到达地球表面的正常阳光量减少了10%。这导致北半球平均气温下降了0.5 ~ 0.6 ℃，全球气温下降约0.4 ℃。

喀拉喀托火山和皮纳图博火山都对气候造成了短期的瞬时效应。这是因为注入大气的二氧化硫和尘埃量相对较低，同时注入的水量也意味着大部分物质在几年内会被冲刷出大气（图28）。然而与超级火山的爆发相比，这两次爆发就不算什么了。超级火山的喷发量比喀拉喀托火山大几千倍。当地球上的岩浆从热点①升入地壳，但无法冲破地壳时，就会发生超级火山爆发。压力在一个巨大且不断增长的岩浆池中逐渐增强，直到地壳无法承受。它们也可以在板块的会聚边界处形成，例如多峇火山，它最后一次爆发是在大约7.4万年前，向大气层注入了

----

① 热点（hotspot），指地球表面长期历经活跃的火山活动的地区。——译注

（a）火山爆发

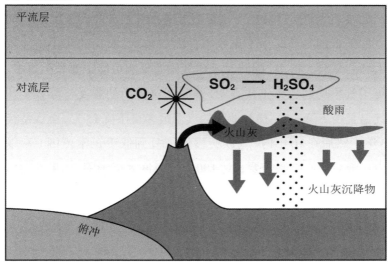

（b）超级火山爆发

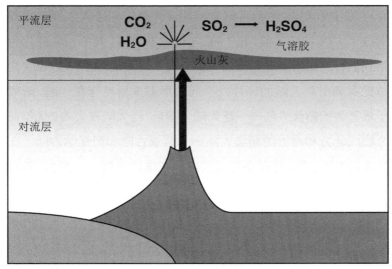

图28　火山爆发对大气成分的影响

74

73

大约2 800立方千米的物质。它们还可以在大陆热点地区形成，例如黄石公园，它最后一次爆发是在210万年前，喷出了2 500立方千米的物质。由于这些事件的规模，二氧化硫和尘埃注入大气的量要大得多，因此对全球气候的影响可能也要长久得多。英国气象局的建模工作表明，热带超级火山的爆发将导致全球气温至少下降6 ℃，热带地区的气温最多可下降15 ℃，至少持续3年。然后在10年内，气候会慢慢恢复到低于正常范围1 ℃以内。最后的影响则需要长达100年的时间才能消除，如果真的出现了这种情况，对我们来说将是毁灭性的。不过从地质学角度来说，这是一个非常短期的事件，对气候系统的长期影响不大。

冰室和温室世界

　　板块构造推动了全球大陆的缓慢移动，从超级大陆变成了零碎的各个大陆，然后再聚拢回超级大陆。超级大陆罗迪尼亚形成于大约11亿年前，在7.5亿年前左右解体，其中一块碎片包括我们目前在南半球发现的大部分大陆。大约3亿年前，板块构造把罗迪尼亚大陆的碎片以不同的形态重新组合在一起，形成了最著名的超级大陆——泛大陆。随后，泛大陆在大约2亿年前分裂成北方和南方的超级大陆——劳亚古陆和冈瓦纳古陆。在过去的1亿年里，这两个超级大陆都在继续分裂。一方面，随着各块大陆的聚拢，冰室气候因此形成。因为海床生成不足，所以海平面很低。大面积超级高原带来了强烈的雨影效应，降雨量减少，气候变得更加凉爽干燥。另一方面，温室气候随着大陆的分散而形成，由于海床展开程度高，海平面也高。海洋断裂带的产生使大气中的二氧化碳含量也相对较高，可能是现在的3倍以

75

上。温暖湿润的气候由此形成。

这些超级大陆的形成和解体对演化产生了巨大的影响。超级大陆对生命极为不利。首先，大陆架的海域数量大幅减少，我们认为多细胞生命可能就是从这些海域里开始的。其次，大陆腹地非常干燥，全球气候通常很冷。一些关键的大规模生物集群灭绝与超级大陆的形成有关。例如，据估计，在2.5亿年前的二叠纪–三叠纪生物集群灭绝中，高达96%的海洋物种和70%的陆生脊椎动物物种灭绝了，该事件被称为"大规模灭绝之母"（图29）。在大约5.5亿年前的寒武纪，当罗迪尼亚超级大陆解体后，复杂的多细胞生物出现了爆炸式增长，也不足为奇。

雪球地球

有一种观点认为，在大约6.5亿年前，地球表面至少曾有一次完全冻结的时期，即所谓的"雪球地球"假说。这可以用来解释在热带地区发现的沉积物，它们显示出冰川的特征，表明热带地区一定有过很多冰。反对这一观点的人认为，地质学上的证据并不表明出现过全球冰冻。此外，要让海洋全都变成冰也有困难，就连覆盖满雪泥也不容易。世界一旦处于雪球状态，也实在难以证明它随后是如何摆脱冰冻状态的。一个答案是，这种情况可以通过大气中二氧化碳和甲烷的缓慢积累，最终达到临界浓度来实现，让大气变暖到足以开始融化过程。还有很多问题没有答案，包括地球是一个**完整的**雪球，还是一个赤道地带有一条狭窄开放水带的"雪泥球"。但特别有趣的是，复杂生命的演化结束了再次出现雪球地球的可能性。布里斯托尔大学的安迪·里奇韦尔教授认为，当前形成方解石壳层的海洋微生物的

76

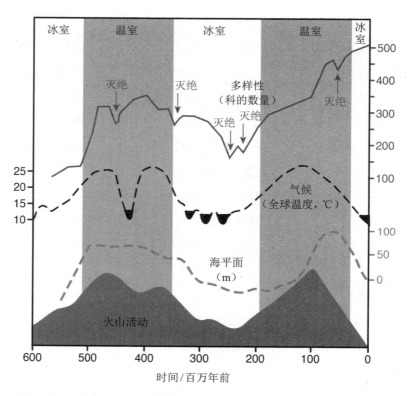

图29 构造、海平面、气候、生物多样性和物种灭绝之间的长期联系

演化极大地缓冲了海洋的碳酸盐系统，以至于现在不可能出现大气中二氧化碳的极端变化，而这种变化是让世界陷入或脱离雪球或雪泥球状态所必需的。

总　结

　　现代气候系统是各大陆在地球表面缓慢运动的产物。我们目前处于一座"冰室世界"中，因为两个极点上都有大陆或在大陆包围之下。大气中二氧化碳的减少使得南极洲和格陵兰岛上的永久性冰盖得以形成。这产生了一个非常强烈的赤道-极地温度梯度，至少有60 ℃的温差，推动了非常活跃的气候系统。目前纵向大陆和海洋通道的排列，在北大西洋和南极洲造就了大量的深层水。现代山脉和高原的位置控制了世界主要沙漠和季风系统的位置。大陆的运动也严重影响了全球和区域的气候，气候又进而影响了演化。当今的气候归根结底是板块构造和大陆随机位置的产物。

78

全球气候变冷

引　言

5 000万年前，地球是一个与现在大不相同的地方。世界既温暖又湿润，雨林从加拿大北部一直延伸到南方的巴塔哥尼亚。地球是如何从郁郁葱葱、生机勃勃演变到今天的冰封凉爽的？导致大冰期开始的原因是什么？如果把5 000万年前的世界地图和今天的进行比较，它们貌似一致，细看才知区别。我们在第五章中看到，各大洲在地球表面的运动非常缓慢，但位置的微小变化却对全球气候产生了深远的影响。在过去的5 000万年里，这些微小的变化使地球的气候从温室世界变成了冰室世界。

过去的1亿年

在过去的1亿年里，南极洲一直坐落在南极上，美洲和亚洲大陆包围着北极。但我们只有在过去的250万年里才在大冰期（即所谓的冰期-间冰期旋回）中循环往复。因此，一定还有额外的因素控制着地球的温度，特别是需要一种手段来冷却极点上或周围的大陆。就南极洲而言，冰层是在3 500万年前才开始形成的（图30）。在此之前，南极洲覆盖着郁郁葱葱的温带森林：在那里发现了恐龙的骨骼，可以追溯到6 500万年前恐龙灭绝之前。3 500万年前发生的变化是小构造运动的高潮。南美洲和澳大利亚慢慢地远离了南极地区。大约3 500万年前，塔斯马尼亚岛和南极洲之间的海洋日渐开阔。随后在大约3 000万年前，南美洲和南极洲之间的德雷克海峡开放，这是最可怕的一片海洋。这使得南大洋开始在南极洲周围循环。南大洋的作用很像家中冰箱里的冷却液。它在围绕南极洲流动时，从南极洲吸收热量，然后释放给与之混合的大西洋、印度洋和太平洋。大陆之间这些看似很小的海洋通道的打通，产生了一片可以完全围绕南极洲循环的海洋，不断吸走大陆的热量。这个过程如此高效，以至于保证了现在的南极洲上有足够多的冰，如果这些冰全部融化，全球海平面将上升逾65米——这个高度足以淹没自由女神像的头部。这种南极冰川化的构造学原因，也是科学家相信全球变暖不会导致东南极冰盖融化的理由——如果冰盖融化，将导致海平面上升约60米。而不稳定的西南极冰盖就不一样了（见第八章）。

　　然而，3 000万年前冰封的南极洲并没有持续多久。在2 500万年前到1 000万年前，南极洲不再完全被冰雪所覆盖。问题是，为什么1 000万年前世界又开始降温？为什么冰层开始在北半球堆积？古气候学家认为，如果要维持一个星球始终寒冷，大气中相对较低的二氧化碳水平是不可或缺的条件。计算机模型显示，如果大气中二氧化碳含量很高，即使有海洋的热交换，也

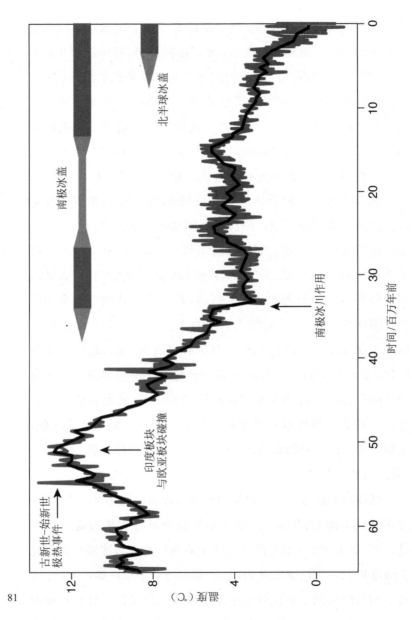

图 30 过去 6 500 万年的全球气候

81

无法让南极洲结冰。那么是什么原因导致二氧化碳水平越来越低,北方又为何开始结冰了呢?

大冰冻期的根源

1988年,比尔·拉迪曼教授和他当时的研究生莫琳·雷莫在拉蒙特-多尔蒂地球观测站写了一篇极有影响力的论文。他们指出,全球降温和北半球冰盖的累积是由喜马拉雅山-青藏高原和谢拉山-科罗拉多高原的抬升造成的。正如我们在第五章中看到的那样,巨大的高原可以改变大气环流,他们认为这使北半球降温,冰雪堆积。然而,他们当时没有意识到,喜马拉雅山的抬升大多发生在早得多的2 000万年前到1 700万年前,因而为时过早,不可能是北方冰雪的直接原因。但莫琳·雷莫随后提出了一个惊人的观点,即这种抬升可能会造成侵蚀的大量增加,在这个过程中会消耗掉大气中的二氧化碳。这是因为在造山时也会产生雨影。由于空气被迫上升翻越山体,山的一侧因此有更多的降雨。这也是山脉比起伏平缓的山丘侵蚀得更快的原因。她认为,这些多出来的雨水和大气中的二氧化碳形成了弱碳酸溶液,从而溶解了岩石。但有趣的是,只有硅酸盐矿物的风化才会对大气中的二氧化碳含量产生影响,因为碳酸对碳酸盐岩石的风化会使二氧化碳回到大气中。由于喜马拉雅山的大部分地区都是由硅酸盐岩石组成的,所以有很多岩石可以锁住大气中的二氧化碳。溶解在雨水中的新矿物就会被冲入海洋,海洋浮游生物利用这些碳酸钙生成了外壳。海洋生物群的方解石质骨骸最终会以深海沉积物的形式储存下来,从而在其所处的海洋地壳生命周期内从全球碳循环中消失。这是一条将大气

82

中的二氧化碳从大气中去除并倾倒在海洋底部的快速通道。大气中二氧化碳长期变化的地质学证据确实支持这样的观点，即在过去的2 000万年里，大气中的二氧化碳含量已显著下降了。科学家们对这一理论的唯一疑问是，到底是什么阻止了这一进程。以西藏地区过去2 000万年被侵蚀的岩石数量来看，大气中所有的二氧化碳应该都被剥离出来了。因此，一定有其他的自然机制帮助维持大气中二氧化碳的平衡，因为大气中二氧化碳的长期浓度是深海中由风化与沉积所清除的量、俯冲带所回收的量和火山所排放的量之间平衡的结果。

在1 000万年至500万年前，随着大气中二氧化碳含量的下降，格陵兰冰盖开始形成。有趣的是，格陵兰岛起初是从南部开始冰川化的，这是因为必须有产生冰的水汽来源。所以在500万年前，南极洲和格陵兰岛就有了巨大的冰盖，与今天的情况大致相同。在北美洲和北欧，巨大的冰盖在250万年前的大冰期才开始消长，但有趣的证据表明，大约在600万年前，这些大冰盖确实开始扩展了。此时在北大西洋、北太平洋和挪威海都发现了来自大陆的岩石碎片，这些冰蚀的碎片随后被冰山倾倒在海上。这似乎是启动大冰期的一次失败尝试，原因可能在于地中海。

盐类大危机

大约600万年前，构造上的渐进变化使得直布罗陀海峡封闭，导致了地中海与大西洋的短暂隔离。在这一隔离期间，地中海几度干涸，形成了大量的蒸发岩（盐类）沉积。想象一下一片巨大版的死海，几米深的海水覆盖了一大片区域。这一全球性的气候事件被称为墨西拿盐度危机，因为世界海洋中近

气候

83

6%的溶解盐被清除了。到了550万年前，地中海已被完全隔离，成为一片盐漠（图31）。这与古气候记录显示的北半球开始冰川化的时间大致相同。但在大约530万年前，直布罗陀海峡重新开放，造成了墨西拿末世洪水，也被称为赞克尔期大洪水或赞克尔期洪灾。科学家们设想有一个巨大的瀑布，比今天委内瑞拉的安赫尔瀑布（979米）还要高，也比阿根廷和巴西边界的伊瓜苏瀑布或加拿大和美国边界的尼亚加拉瀑布都要壮

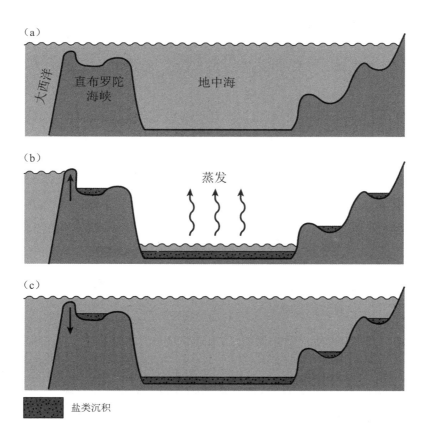

盐类沉积

图31　大约500万年前的地中海墨西拿期"盐度危机"和"末世洪水"

阔得多。最近对直布罗陀海峡地下结构的研究表明,洪泛通道可能以一种比较渐进的方式流入干涸的地中海。洪水可能发生了几个月或几年,这意味着大量溶解的盐分通过地中海-大西洋通道被输回了大洋。这阻止了大冰期的发展,而且完全是由于海洋的循环方式。正如我们在第二章中看到的那样,墨西哥湾流不仅使欧洲保持温暖,还推动了深海的循环,让整个地球保持相对的温暖。500万年前,深海环流还没有如今这般强大。这是因为盐度较低的太平洋海水还能通过巴拿马海洋通道渗漏出来,下文将会讨论这个问题。由于墨西拿末世洪水导致盐分突然大幅增加,提高了北大西洋的盐度,因此确保了墨西哥湾流的异常活跃以及北欧海域的海水下沉。大约500万年前,所有这些热带的热量被有效地输送到北方,阻止了地球进一步滑入大冰期。我们不得不再等250万年,全球气候才能

准备好再次去尝试。

巴拿马悖论

　　另一个重要的构造控制因素是太平洋-加勒比海通道的关闭,地质学家认为它是造成大冰期的一个诱因。如今在苏黎世大学和阿尔弗雷德·魏格纳研究所的两位教授格拉尔德·豪格和拉尔夫·蒂德曼利用海洋沉积物的证据表明,巴拿马海洋通道在450万年前开始关闭,并在大约200万年前最终关闭。然而,巴拿马通道的关闭造成了一个悖论,因为它既会促成也会阻碍大冰期的开始。首先,流入加勒比海的太平洋表层水减少,提高了加勒比海的盐度,因为太平洋的海水比北大西洋的更淡。这将提高由墨西哥湾流和北大西洋洋流向北输送的水的盐度,

正如我们在上文看到的那样,如此便会促进深层水的形成。墨西哥湾流的强度增加和深层水的形成会对大冰期的开始产生不利影响,因为这加强了海洋热量向高纬度地区的输送,阻碍了冰盖的形成。所以在大约500万年前启动大冰期的尝试失败后,巴拿马海洋通道的逐步关闭继续增加了北上的热传导,牵制了寒流。但这里有一个悖论:大型冰盖的形成需要两个条件——低温和大量的水分。增强的墨西哥湾流也向北方输送了更多的水分,冰盖的形成已万事俱备。这意味着北半球大型冰盖的形成可以在温度较高的时候开始,因为所有额外的水分都被输送到北方,随时会以雪的形式落下,形成冰盖。

为什么是在250万年前?

单纯的构造作用力不能解释北半球冰川作用速度惊人的加剧(图32)。我利用海洋沉积物进行的工作表明,在向大冰期过渡的过程中,有三个主要步骤。证据基于岩石碎片被冰从大陆上扯落的时间,而那些碎片是由冰山沉积在邻近的海洋盆地的。首先,大约在274万年前,欧亚大陆北极地区和东北亚地区的冰盖开始扩展,东北美洲冰盖的扩展也有一定的证据;其次,在270万年前,阿拉斯加开始出现冰盖;最后,最大的冰盖出现在东北美洲大陆上,在254万年前达到最大的面积。因此,在不到20万年的时间里,我们就从被基尔大学的迈克尔·萨恩泰因 86 教授称为"气候的黄金时代"的上新世早期温暖、温和的气候进入了大冰期。

北半球冰川作用开始加剧的时机肯定另有原因。有人认为,轨道驱动的变化(地球绕太阳公转方式的变化)可能是造 87

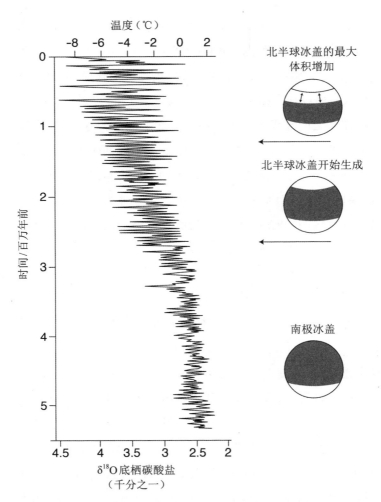

温度（℃）

北半球冰盖的最大
体积增加

北半球冰盖开始生成

南极冰盖

δ¹⁸O底栖碳酸盐
（千分之一）

$\delta^{18}O$底栖碳酸盐
（千分之一）

图32　过去500万年的全球气候

成全球降温的一个重要机制。下一章将讨论地球多次摆动的
细节，以及它们如何导致各冰期的消长。虽然这些摆动各自都
是以数万年为单位的，但也有时间长得多的偏差。例如其中最
重要的参数之一是倾角或斜度，也就是地球自转轴的上下摆

动——或者换一种说法，就是地球自转轴相对于其轨道平面的倾斜度。在 4.1 万年的时间里，地球的自转轴会向太阳倾斜得多一点，然后再少一点。这个变化并不大，从 21.8° 到 24.4° 不等。在第一章中，我们探讨了自转轴的斜度如何造就了季节更替。因此，斜度越大，夏冬两季之间的差异就越大。倾斜的幅度在 125 万年里发生了改变。在 500 万年前和 250 万年前，地球两次试图以冰雪覆盖北半球，斜度的变化都增加到了最大值。这使得每个季节的变化非常明显，最重要的是北方寒冷的夏天保证了冰层不会融化，让冰层可以发展成冰盖。

热带地区对冰期的反作用

北半球冰川作用加剧的开始并不止影响了高纬度地区。在大冰期开始的 50 万年后，热带地区的情况似乎发生了变化。在 200 万年前，太平洋似乎有一个非常微弱的东西向海面温度梯度，但后来这个梯度越来越大，显示出热带和亚热带向现代环流模式的转换，沃克环流相对较强，亚热带温度较低。沃克环流是哈德来环流东西向的大气组成部分，对控制热带地区的降雨起着重要作用。沃克环流也是恩索（ENSO，见第三章）的关键因素。因此，在 200 万年前，由于当时的沃克环流相对较弱，恩索可能尚未以当今的形式存在。沃克环流的发展似乎也与人类的早期演化有关。东西向环流的加强似乎在东非大裂谷中产生了深邃但短暂存在的淡水湖泊。最近有人推测，这种独特的气候节拍以及湖泊在大约 200 万年间的迅速出现和消失，可能与直立人的演化有关。在此期间，直立人的脑容量增加了 80% 以上，我们的祖先也第一次走出了非洲。

88

中更新世气候转型期

中更新世气候转型期（MPT）是指过去80万年以来的某个时候发生的冰期-间冰期旋回的明显延长和加剧（图32）。在中更新世气候转型期之前,冰期-间冰期旋回似乎每隔4.1万年发生一次,与地球斜度的缓慢变化相对应。在过去80万年以来,冰期-间冰期旋回好像要长得多,平均超过10万年。这些旋回的特点也发生了变化。在中更新世气候转型期之前,冰期和间冰期之间过渡平稳,世界似乎在每种气候中度过的时间也是相等的。在中更新世气候转型期之后,这些旋回变成了锯齿状,冰层在8万年的时间里堆积起来,使地球深陷于强烈的冰期,随后冰层又迅速消退,在4 000年内全部消失。然后,气候在类似于当前气候的间冰期停留了大约1万年,接下来又回到了冰期。对于这种锯齿状的模式,有一种说法是,庞大得多的冰盖非常不稳定,因此只要气候稍有变化,它们就会迅速崩塌,整个气候系统就会回升到间冰期。在下一章,我们将更详细地研究这些最近的冰期-间冰期旋回。

第七章

大冰期

引　言

　　1658年，英国阿马的厄谢尔大主教观察了他周围的景观特征，将其成因归结为诺亚大洪水。他利用《圣经》，煞费苦心地将洪水和景观的年代确定为公元前4004年。直到1787年，日内瓦贵族、物理学家、阿尔卑斯山旅行家奥拉斯-贝内迪克特·德索绪尔才认识到，阿尔卑斯山的漂砾已被移到数百英里之外汝拉山脉的山坡下，并推断出山地冰川一定可以回溯到更久远的古代。这一发现不得不等到1837年，瑞士地质学家路易斯·阿加西才根据漂砾和终碛的证据，提出了他的"大冰期"或"冰川"理论。终碛是由侵蚀的沉积物被冰盖像推土机一样不断推进而形成的山丘。当冰盖达到最大限度时，沉积物就会淤积成一排小丘，描摹出冰盖的前缘。1909年，德国和奥地利的地理学家阿尔布雷希特·彭克和爱德华·勃吕克纳出版了一部名为《大冰期的阿尔卑斯山》的三卷本著作。他们得出的结论是，阿尔卑斯山有四次主要的冰期，即贡兹期、明德期、里斯期和玉木期。地面或陆上证据的缺点是不连续，证据可能会被后来的冰盖推进破坏。因此，直到1960年代，人们从海洋底部找到了绵长且连续的沉积岩心，才知道曾经有多少次冰期。我们现在有能力在海洋中钻到4英里的深度，还能打捞出海床下半英里以上的沉积物。通过研究这些海洋沉积物，科学家们已经记录了过去250万年中发生的50次冰期。

90

大冰期的消长

我们现在知道，冰期-间冰期旋回是第四纪，即过去250万年以来的基本特征。巨大的大陆冰盖的消长变化是由地球绕太阳轨道的变化引发的。地球长期以来都在其轴线上摆动，从而改变了地球不同部分所接收的阳光或太阳能的量。这些微小的变化足以推动或迫使气候发生变化。然而，这些冰盖的消长并不由地球轨道的摆动引起，而是由地球气候的反应所造成的，它将区域太阳能相对较小的变化转化为重大的气候变化。例如，地球当今的位置与2.1万年前最后一次冰期时所在的位置非常接近。所以控制气候的不是轨道的准确位置，而是轨道位置的变化。有三个主要的轨道参数或摆动，分别称为偏心率、倾角或斜度，以及进动（方框4）。从方框中可以看出，每个参数都有一个独特的周期及其对气候的影响。但更令人兴奋的是，我们可以将这些参数结合起来，看看它们是如何导致气候进入或脱离大冰期的。

机械式的气候？

结合全部三个轨道参数的影响，可以计算出任意纬度在过

气
候

90

去接收到的太阳能。塞尔维亚杰出的数学家和气候学家米卢
廷·米兰科维奇在1949年提出,北纬65°,也就是北极圈以南
的夏季日射对控制冰期−间冰期旋回至关重要(图34)。他认
为,如果夏季日射减少得足够多,那么冰块就可以在夏季留存
下来,并开始堆积,最终形成冰盖。轨道驱动确实对夏季日射
有很大的影响;过去50万年中太阳辐射的最大变化,相当于把
今天在北纬65°处接受的夏季辐射量减少到现在其北逾550千
米外的北纬77°所接受的辐射量。简单来说,这使得目前挪威

方框4 轨道驱动

主要的轨道参数或摆动有三个:偏心率、倾角(斜度),
以及进动(图33),它们对地球的长期气候有重大的影响。

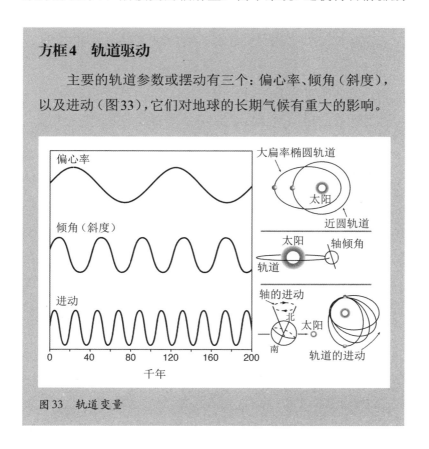

图33 轨道变量

偏心率是指地球围绕太阳运行的轨道形状,从圆(进动)到椭圆不等。这些变化发生的周期约为9.6万年,另外还有一个约40万年的长周期。换一种说法,椭圆的长轴随着时间的变化而变化。在近代,每年1月3日地球离太阳最近,约1.46亿千米,这个位置被称为近日点。7月4日,地球与太阳的距离最远,约为1.56亿千米,这个位置被称为远日点。偏心率的变化只会引起日射总量的微小变化,但如果与进动相结合,则会产生显著的季节性影响。如果地球的轨道是正圆形的,日射量就不会有季节性的变化。如今,地球在近日点接受的平均辐射量约为每平方米351瓦特,远日点为每平方米329瓦特,差异约为6%。但在过去500万年中偏心率(椭圆长度)最大时,差异可能大至30%。米卢廷·米兰科维奇在1949年提出,当夏季太阳距离较远时,北方的冰盖更容易形成,因此每年都有一些去年冬天的雪留存下来。偏心率的另一个作用是调节进动的影响。但必须指出,偏心率是到目前为止所有三个轨道参数中影响最弱的一个。

地球自转轴相对于其轨道平面的倾角或斜度在4.1万年里均介于21.8°和24.4°之间。如第一章所述,正是自转轴的斜度导致了季节的变化。倾角越大,夏季和冬季接收到的日射差异就越大。米卢廷·米兰科维奇认为,北半球夏季越冷,降雪的可能性就越大,这导致了冰川和冰盖的逐渐形成。

进动包括与地球的椭圆轨道（偏心率）及其自转轴有关的部分。地球的自转轴每隔2.7万年就会发生一次进动，这一过程类似于玩具陀螺转轴的旋转。进动会导致任何特定日期的地日距离发生变化，例如北半球夏季开始的日期。正是由于轨道参数的不同组合，导致了2.3万年和1.9万年这两种不同的进动周期。将自转轴的进动与轨道的进动变化结合起来，就会产生2.3万年的周期。但是，偏心率（变化周期9.6万年）和自转轴的进动相结合，也会产生1.9万年的周期。这两个周期结合在一起，两个半球的近日点因而平均每隔2.17万年就会与夏季相吻合。进动的影响在热带地区最为显著（相比之下，赤道受倾角的影响为零）。因此，虽然倾角明显影响了高纬度地区气候的变化，而高纬度地区气候的变化可能最终影响了热带地区，但日射对热带地区的直接影响仅仅是由偏心率调节的进动所造成的。

将偏心率、倾角和进动的影响结合起来，就可以计算出古往今来任何纬度的日射量。图34显示了北纬65°的计算日射量与过去60万年全球海平面变化所代表的冰盖面积变化的比较。

中部的冰区边界降到了苏格兰中部的纬度。北纬65°日射的这些减少是由于偏心率拉长了夏季的地日距离，倾角很小，而进动使夏季处于偏心率产生的最大地日距离。之所以是北纬65°而不是南纬65°在控制气候，原因很简单。堆积在北半球的冰

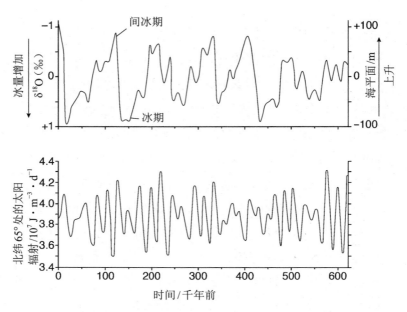

图 34 北半球太阳辐射和全球海平面的比较

层都有许多大陆可以依托。相反,在南半球,冰层的增长受到
94 南大洋的限制,因为南极洲上任何新增的冰都会落入大洋,被
卷向温暖的海洋(图35)。所以关于冰川的传统观点是,北半
球温带夏季太阳能量低,使得冰能够在夏季留存,因此冰盖开
始在北方大陆上堆积。但这种看似简单的机械世界观其实要
复杂得多,因为轨道变化对季节的影响非常小,是气候系统的
反馈放大了这些变化。

导致冰期-间冰期旋回的原因

轨道驱动本身并不足以驱动上文提到的冰期-间冰期的气
候变化。相反,地球系统通过各种反馈机制,放大和转化了地球

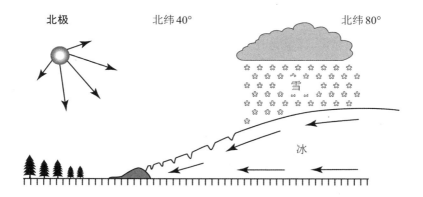

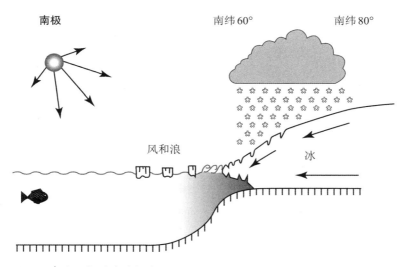

图35 南北两极的冰盖扩展

表面接收到的太阳能的变化。我们先来看看冰期的形成。　95

　　首先需要发生的是夏季气温略微下降。由于夏季温度的这种变化，冰雪不断积累，反照，即反射回空间的阳光就会增加。把更多的阳光反射回太空的过程会抑制局部的温度，从而促进更多冰雪的积累，进一步增加该地区的反照率，产生所谓的"冰–

反照率"反馈。所以一旦有了小冰盖,就会改变周围的环境,冰雪会越来越多,冰盖也会越来越大。

当冰盖大到足以干扰大气中的行星波时,就会引发另一种反馈(见第五章,图27),北美的劳伦泰德冰盖尤其如此。这改变了横跨北大西洋的风暴路径,阻止了墨西哥湾流和北大西洋漂流涌向如今这样靠北的地方。这种表层海洋的变化,再加上北欧海域和大西洋因大面积大陆冰盖的存在而导致的融水普遍增加,最终造成深层水形成的减少。格陵兰海和拉布拉多海的深层水形成是现代气候的重要特征。通过减少深层水的形成,被拉向北方的暖水量降低,所有这些都导致北半球的降温加剧和冰盖的扩展。

对于上述的地文气候反馈和大气中温室气体的作用,古气候学家目前仍在争论不休。极地冰中的气泡表明,在每个冰期,二氧化碳都会减少三分之一,甲烷则会减少一半。这些变化加剧了每个冰期发生的降温,有助于形成更多的冰。因此,争论还在继续:地球轨道的变化是否会影响温室气体的产生,使地球降温,让北半球的大陆容易形成大面积的冰盖?还是地球轨道的变化启动了北半球大冰盖的形成,继而改变了全球气候,减少了温室气体的产生,从而延长和加剧了冰期?很抱歉,这些问题还没有定论。然而我们知道,温室气体在冰期-间冰期旋回中发挥了关键作用。我们还知道,温室气体浓度的变化始终先于全球气温的变化。

有一个很重要的问题,即这些反馈为什么不会失控,进而冻结整个地球?答案是一个叫作"水分限制"的过程阻止了失控效应。形成冰盖需要气候寒冷湿润。不过,随着温暖的地表水被迫进一步流向南方,形成冰盖所需的水分供给减少。所以通过改变大气和海洋环流,冰盖最终失去了水分。

过去100万年来,冰盖的形成需要8万年才能达到冰面的最大范围。所以上一次出现这种情况是在2.1万年前。然而,冰消雪融的速度要快得多。这个过程被称为"冰川减退",通常最多只需要4 000年。夏季北纬65°左右地区接收到的太阳能增加,就会引发这种冰川减退的现象。这促使北半球的冰盖轻微消融。大气中二氧化碳和甲烷的增加加速全球变暖,并助长了大陆上大型冰盖的融化。但这些过程必须与冰盖的反照效应相抗衡,而反照效应所产生的小气候则能有效地保持冰盖的完整。导致冰雪迅速消融的原因是冰盖融化所引起的海平面上升,与海洋相邻的大型冰盖会被上升的海平面削凿。海水的温度最低可降到-1.8 ℃左右,而冰盖底部的温度通常低于-30 ℃。水对冰的影响类似于把热水放在一盆冰淇淋的下面。对冰盖的削凿会导致更多的融化和冰层坍塌入海,这反过来又提高了海平面,引发更多的削凿。海平面的这种反馈过程发生得非常快。一旦 98 冰盖全线退却,那么上面讨论的其他反馈机制就会倒转过来。

剖析最近一次冰期

如果我们把注意力集中在距今仅2.1万年前的最近一次冰期,就可以看到冰川对地球气候的影响有多大。在北美洲,几乎连续的冰层横跨了从太平洋到大西洋的整个大陆。它由两个独立的冰盖组成:东部以哈德孙湾为中心的劳伦泰德冰盖,以及西部海岸山脉和落基山脉的科迪勒拉冰盖。劳伦泰德冰盖覆盖了逾1 300万平方千米的土地,在哈德孙湾最深处达到了超过

3 300米的厚度。它的最大范围从纽约延伸到了辛辛那提、圣路易斯和卡尔加里。欧洲有两个主要的冰盖，即芬诺-斯堪的纳维亚冰盖和英国冰盖，在欧洲阿尔卑斯山上还有一个小冰盖。英国冰盖在多个冰期与斯堪的纳维亚冰盖合并。每个冰期的平均面积约为34万平方千米。在最后一个冰期，冰盖南下到达不列颠群岛的中部，直抵诺福克郡的边缘。芬诺-斯堪的纳维亚冰盖比英国冰盖大得多，面积达660万平方千米，从挪威一直延伸到俄罗斯的乌拉尔山脉。我们也不能忘记南半球，因为在巴塔哥尼亚、南非、澳大利亚南部和新西兰都有大型冰盖。此外，南极的冰盖扩展了约10%，季节性的海冰又在大陆之外的地方延伸了500英里。这些冰盖冻结的水量之巨难以想象。对海洋研究一番不失为一种理解的方法。海洋覆盖了地球70%以上的面积，如此大量的水被从海洋中吸走并冻结在冰盖中，以至于海平面下降了逾120米，大约相当于"伦敦眼"的高度。如果今天南极洲和格陵兰岛上的冰全部融化，海平面将会上升70米。在冰期，全球气温比如今低5～6℃，但这种气温的下降并不均匀，高纬度地区降温幅度高达12℃，而即使是热带地区也会降温2～5℃。冰期也非常干燥，大气中含有大量的灰尘。例如，在中国北部、美国东部、中东欧、中亚和巴塔哥尼亚等地，都有数百米厚的尘埃沉积，称为"黄土沉积"，这些尘埃都是在冰期积累起来的。

冰雪塑造了陆地

在冰期，这些巨大冰盖的存在深刻影响了当地的气候、植被和景观。高纬度地区的大片北方森林遭到了彻底破坏，因为它们曾经占据的土地被不断扩展的冰盖所覆盖。大气中水分的降低大大减少了降雨量，世界上的大型湿地和热带雨林也随之萎缩。巨大的大陆冰盖对景观也产生了深远的影响。在温带地区，很少有地方没有受到过冰期的侵袭。行走在北欧和北美，冰期对景观的巨大影响随处可见。这些影响为电影提供了很好的布景，比如《魔戒》三部曲就是在新西兰拍摄的，而人们所看到的山峦荒野等极致景观就是冰盖在岛屿上磨砺了数千年的结果。所以，下次再看这些电影的时候，想想"冰盖"吧。冰盖留下了U形谷、峡湾、冰碛，以及被称为"鼓丘"的蛋形土墩。就连泰晤士河现在的位置也是冰造成的。泰晤士河以前流经伦敦北部的圣奥尔本斯，在埃塞克斯郡汇入北海。欧洲的倒数第二次冰期非常强烈，冰盖使其南下到了伦敦北部，把泰晤士河的河道改到了现在的路径。所以伦敦的地理环境主要受冰期控制。美国很多主要河流的河道都经历过改动，要么是由于某个冰盖的位置，要么是由于1.2万年前冰盖融化时突然出现的大量融水。劳伦泰德河和密西西比河的河道就是上一个冰期结束时这些大洪水的遗迹。

　　由于全球海平面降低了120米，地球的地理环境也发生了变化，这意味着大陆改变了形状。像英国这样的岛屿成为大陆的一部分，也就是说在上一个冰期，人们有可能步行通过英吉利海峡去法国。唯一能阻止你的是一条巨大的新河，它从现在的英吉利海峡中心流过，把泰晤士河、莱茵河和塞纳河的水带到大西洋去。海平面的降低在世界各地形成了陆桥，新的物种得以入侵新的地区。斯里兰卡、日本、英国、西西里岛、巴布亚新几内亚和福克兰群岛等世界各地的岛屿都成为相邻大陆的一部分。

例如,将东北亚和阿拉斯加隔开、横跨白令海的岛链连在了一起。因此,在上一个冰期结束时,随着气候开始变暖,人类第一次从亚洲跨进北美洲,移居新世界。

亚马孙地区消失的草地

冰期显然影响到全球气候系统,但关于冰期对热带地区的影响却存在争议。地球表面的一半位于南北回归线之间的热带地区,包括世界上全部的热带雨林。其中就面积和物种多样性而言,最重要的区域是亚马孙地区。亚马孙盆地是世界上最大的盆地,面积达700万平方千米,那里排放的淡水约占流入海洋总量的20%。该盆地的绝大部分被极其多样的雨林所覆盖。

1969年,哈费尔[①]提出了一个奇妙的理论,指出亚马孙地区如此丰富的物种多样性与冰期有关。他认为,在每一个冰期,热带地区较低的温度和较少的降水量使得热带雨林大多被稀树草原所取代。然而,一些热带雨林会在小型"避难所",即被草原包围的雨林孤岛中幸存下来。这些孤立的雨林地块会成为演化的温床,产生许多新的物种。在每一个冰期结束时,这些零星的雨林又重新合在一处,物种多样性和地方特有现象均胜于以往。然而到了1990年代末,这一理论受到了攻击,因为越来越多的科学家没有发现稀树草原的大量增加。我们现在从孢粉记录和计算机模型中得知,在亚马孙地区,干燥和寒冷的条件结合在一起,意味着稀树草原确实侵占了一部分边缘地区,亚马孙雨林的面

---

① 于尔根·哈费尔(1932—2010),德国鸟类学家、生物地理学家和地质学家。最令人印象深刻的是他关于更新世期间亚马孙森林隐蔽地的理论,该理论有助于研究生物群的标本化和多样化。——译注

积因而减少到如今面积的80%。但亚马孙雨林能够度过冰期甚至蓬勃发展，证明了热带雨林在全球生态系统中的适应力和重要性。雨林之所以能在冰期幸存下来，原因之一是寒冷的环境实际上有助于缓解雨量降低的问题：低温减少了树木的蒸发量，从而削弱了对雨林减少至关重要的水分流失。然而亚马孙雨林的物种构成在冰期发生了重大变化。例如，我们从孢粉记录中知道，目前在安第斯山脉发现的许多树种都曾生长在亚马孙森林的中心地带。这是由于在温暖的间冰期，如今那些更适应寒冷的物种被推升至更高也"更冷"的高地。这一点很重要，因为在过去的100万年里，地球有大约80%的时间处于冰川气候，所以这意味着我们不能把现在的亚马孙雨林看成正常状况。因此，在上一个冰期重建的亚马孙森林中，安第斯和低地热带树种以及常绿和半常绿树种的多样化组合是常态。冰期的亚马孙地区缺乏草原，还意味着我们要为亚马孙雨林丰富的物种多样性寻找其他的演化机制，冰期可能并非其原因。

不稳定的冰期

从很多方面来说，冰期确实应该被称为"气候过山车"，因为冰盖天生不稳定，在冰期，随着冰盖的突然崩塌和再次形成，气候会从一种状态急剧转变到另一种状态。大多数的变化发生在千年的时间尺度上，但这些极端事件的开始可以在短短三年内发生。这些事件中最令人印象深刻的是海因里希事件。海因里希事件是由拉蒙特-多尔蒂地球观测站的古海洋学家沃利·布勒克尔教授以哈特穆特·海因里希的名字命名的，后者曾在1988年撰写一篇论文对其加以描述。海因里希事件是指北美劳伦泰德冰盖的大规模崩塌，导致数百万吨的冰块涌入北大西洋。沃利·布勒克尔将它们描述为从北美漂流到欧洲的冰山舰队。在这些硕大的冰山搁浅的法国北部海岸，人们发现了巨大的凿痕。这些海因里希事件是在冰川气候不稳定的大背景下发生的，是北大西洋地区周边最极端的冰川状况的短暂表现。在格陵兰冰芯记录中，海因里希事件明显表现为本已寒冷的冰川气候温度进一步下降了 2～3 ℃。海因里希事件产生了全球性的影响，远在南美洲、北太平洋、圣巴巴拉盆地、阿拉伯海、中国南海和日本海等地都有重大气候变化的证据。在北大西洋地区周边的这些事件中，北美洲和欧洲都出现了更寒冷的情况。在北大西洋，融化的冰山数量众多，增加了大量寒冷的淡水，海面的温度和盐度降低到表层水无法下沉的程度。这阻止了北大西洋所有深层水的形成，切断了全球海洋传送带。

　　从大西洋中部的海洋沉积岩岩心中很容易观察到海因里希事件发生的证据。这是因为冰山将大量的岩石带入海洋，当冰山融化时，会在洋底散布岩石碎片的痕迹。通过识别海洋沉积物中的这些事件，并对沉积物中的化石进行年代测定，我们得知在上一个冰期，海因里希事件似乎平均每 7 000 年发生一次。此外，在这些岩石碎片下面还发现了海生蠕虫的小洞穴。这些洞穴通常是看不到的，因为沉积物被前来进食的其他动物混在一起了。要想保存这些管状化石和洞穴，冰山融化带来的落雨般的岩石碎片必须在 3 年内出现，而且速度要快到足以防止其他动物到达沉积物。这些证据表明，北美冰盖的崩塌极其迅速，冰山在不到 3 年的时间里就涌进了大西洋。因此在一个冰期内，从有巨大冰盖的寒冷气候到北美冰盖的部分崩塌所带来的极度寒冷气候，环境状况各不相同。

我们现在知道，在大规模的海因里希事件之间，大约每隔1 500年就会发生一些较小的事件，这些事件被称为D-O事件或循环。有一种说法是，海因里希事件实际上只是超级D-O事件。海因里希事件与D-O事件的巨大区别在于，海因里希事件只出现在冰期，而D-O事件既会出现在间冰期，也会出现在冰期。

海因里希事件的起因

海因里希事件之所以令人着迷，是因为我们可以理解它们发生的时间尺度，而且它们对冰期的气候产生了巨大且深远的影响。因此，对于它们的成因有很多不同的理论。一位名叫道格·麦克阿耶尔的冰川学家提出，海因里希事件的冰山波动是由于劳伦泰德冰盖内部的不稳定性所致。该冰盖位于一片柔软松散的沉积物地层之上；这种沉积物在冻结时不会变形，而是表现得像混凝土一样，所以可以支撑冰盖不断增长的质量。冰盖扩展时，地壳内的地热和冰块移动所释放的摩擦热被上覆冰层的隔热作用困住。这种"羽绒被"效应使得沉积物的温度不断上升，直至达到解冻的临界点。冰层解冻时，沉积物变软，从而润滑了冰盖的底部，导致大量冰块通过哈德孙海峡流入北大西洋。这又会导致冰块的突然流失，从而降低保温效果，造成基底冰层和沉积物地层重新冻结，这时冰层又会恢复到较慢的形成和外移状态。道格·麦克阿耶尔称这是一种暴食狂泻模式，认为所有的冰盖都有自己的不稳定时间，因此芬诺-斯堪的纳维亚、格陵兰岛和冰岛的冰盖会有不同周期的波动。

另一个激动人心的理论是"两极气候跷跷板"的观点——

这是沃利·布勒克尔教授创造的另一个美妙术语。这个理论基于格陵兰岛和南极的冰芯新证据，这些证据表明，在海因里希事件期间，南北两个半球的气候是反相的。所以当北半球在降温时，南极洲却在升温。有人认为，这种所谓的两极气候跷跷板效应可以用南北大西洋的冰盖崩塌和由此产生的融水事件交替出现来解释。每一次融水事件都会改变两个半球深层水形成的相对量，以及由此产生的半球间热量掠夺的方向。目前北半球从南半球吸收热量，以维持墨西哥湾流和北欧海域相对温暖的深层水形成。热量通过从北大西洋流向南大西洋的深层水缓慢回流。所以两极气候的跷跷板模式表明，如果北大西洋周边的冰盖崩塌并将大量的冰山送入海洋，它们就会融化。这些融化的水会让海洋变得过淡，以至于这些表层水都无法下沉。这阻止了北大西洋深层水的形成，北半球也就不再从南半球吸收热量了。如此一来，南半球就会慢慢变暖。也许在 1 000 年后，这种热量的积累足以使南极洲的边缘坍塌，继而阻断南极洲周边的深层水形成，将整个系统逆转过来。这个理论的好处是它也可以解释间冰期，正如我们在上文看到的那样，D–O 循环约为 1 500 年，在冰期和间冰期都会出现。

全新世

我们在大约 1 万年前度过了最后一个冰期，目前正处于被称为全新世的间冰期。间冰期的气候并不是恒定不变的，全新世初期可能比 20 世纪更加温暖湿润。在整个全新世期间，发生过千年尺度的气候事件，即 D–O 循环，导致 2 ℃ 的局部降温。这些事件可能对古典文明产生了重大影响。例如，大约 4 200 年前的寒冷干旱事件与许多古典文明的消亡相吻合，包括埃及的古王国，美索不达米亚的阿卡德帝国，安纳托利亚、希腊和以色列的早期青铜时代社会，印度的印度河流域文明，阿富汗的赫尔曼德文明，以及中国的红山文化。这些千年气候周期中的最后一次是小冰期。该事件实际上包括两个寒冷期，第一个寒冷期是在1 000年前结束的中世纪暖期之后，通常被称为中世纪冷期。中世纪冷期起到了毁灭格陵兰岛上的挪威殖民地的作用，并造成了欧洲的饥荒和大规模移民。它在公元1200年以前逐步开始，到公元1650年左右结束。第二个冷期在传统上被称为小冰期，这可能是全新世晚期北大西洋地区变化最大最迅速的时期，冰芯和深海沉积物记录都表明了这一点。英国的气温平均下降了1 ℃——尽管大家看到在冰天雪地的泰晤士河冰雪集会上创作的美丽画作，都以为气温下降的幅度要大得多。但这是一个神话，因为现在英国的天气要想让泰晤士河冷到结冰几乎不可能。由于1831年拆除了旧伦敦桥，1870年代的泰晤士河拉直工程修建了堤岸（这样伦敦人就可以像巴黎人一样漫步河畔），还有疏浚河道使其成为大英帝国中心的国际港口，泰晤士河早已不再是一条缓慢蜿蜒的河流了。

如果我们考察全球各地的记录，小冰期和中世纪暖期似乎只发生在欧洲北部、美洲东北部和格陵兰岛。因此小冰期是由墨西哥湾流的微小变化和冰岛北部的深层水形成所驱动的区域性气候波动。许多否认气候变化的人认为，全球变暖只是世界从小冰期中恢复过来。但由于世界上大部分地区从未发生过小冰期，所以也就没什么可以恢复的。过去2 000年来的全球温度记录的重建至关重要，因为它们为过去150年的仪器温度数据提

供了背景。它清楚地表明,20世纪和21世纪的气温比过去2 000

年的其他任何时间都要高。

总　结

在过去的250万年里,地球的气候一直被大冰盖的来来去去所主导。这些冰盖是如此之厚,以至于在不过2.1万年前,北美和北欧还堆积着2英里厚的冰层。全球气候发生了根本的变化。冰期的全球平均气温比现在低6 ℃,全球海平面比如今低120米,大气中的二氧化碳含量低三分之一,甲烷也少一半。陆地上所有植物的总生物量少了一半之多。由于这些巨大冰盖的侵蚀和淤沙沉积,地球的景观发生了巨大的变化。大河改道,沧海桑田。随着海洋降低,连接相邻各大陆的陆桥出现了,物种得以在新的土地上拓殖。在过去的250万年里,地球的气候系统似乎也更乐意处于寒冷的状态,而不是如今的温暖状态。

气
候

第八章

未来的气候变化

引　言

　　未来的气候变化与减贫、环境恶化和全球安全一样，是21世纪的决定性挑战之一。问题在于"气候变化"不再仅仅是一个科学问题，它如今还涉及经济、社会学、地缘政治、国家和地方政治、法律、健康等诸多方面。本章将简要探讨什么是"人为的"气候变化，以及全球气候系统开始变化的证据。本章还将解释气候系统的变化为什么会导致不可预测的天气模式和极端天气事件的增加，如风暴、洪水、热浪和干旱。详见我为"牛津通识读本"系列所写的另一本书：《全球变暖》。

人类造成的气候变化

　　强有力的证据表明，我们一直在改变大气中的温室气体含量。对大气中二氧化碳浓度的第一次直接测量始于1958年，测量位置在夏威夷冒纳罗亚火山顶约4 000米的高度，那是一个没有当地污染的偏远地点。而再往前看,我们还分析过格陵兰岛和南极冰盖上冰块中夹带的气泡。这些长期的冰芯记录表明,工业化前的二氧化碳浓度约为体积占比的百万分之二百八十(280 ppmv)。1958年,浓度已达到316 ppmv,并且逐年攀升,2013年6月达到400 ppmv。我们在一个世纪内造成的污染,比贯穿大冰期自然起伏的数千世纪发生的污染还要多。不幸的是,大气中二氧化碳的这种增加在我们目前产生的污染中仅占半数:大约有四分之一的二氧化碳被海洋吸收,还有四分之一被陆地生物圈吸收了。科学家们最担心的一个问题是,大自然提供的这种服务在未来可能会减少,到时情况会变得更糟。

根据联合国政府间气候变化专门委员会(IPCC)2007年的科学报告,过去150年所有温室气体的增加(见第二章)已经大大改变了气候,包括:全球气温平均上升0.75 ℃;海平面上升逾22厘米;降水的季节性和强度发生显著变化,改变了天气模式;

北极海冰和几乎所有大陆冰川大幅退缩。根据美国国家航空航天局(NASA)、美国国家海洋和大气管理局(NOAA)、英国气象局和日本气象厅的数据,在过去150年中,最近十年是有记录以来最温暖的十年(图36)。IPCC在2007年表示,气候的变化有明确证据,有很大把握认为这是人类活动的结果。这一观点得到了众多专家组织的支持,包括英国皇家学会和美国科学促进会。

"重量级证据"

理解未来的气候变化就是要理解科学如何运作,以及"重量级证据"的原则。科学通过利用详细的观察和实验来不断检验想法与理论,从而实现发展。在过去的30年里,气候变化理论一

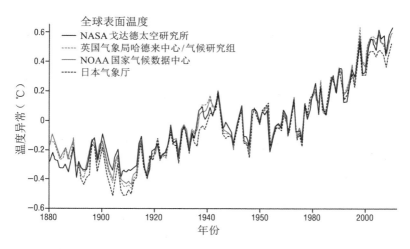

图36　过去120年全球平均地表温度

定是科学中经过最全面检验的观点之一。第一，如前所述，我们已经追踪到了大气中温室气体的上升。第二，我们从实验室和大气测量中得知，这些气体确实会吸收热量。第三，我们跟踪到了20世纪全球气温和海平面上升的重大变化。第四，我们分析了地球系统中与气候有关的物理变化，包括北极和南极洲周围的海冰退缩，各大洲的山地冰川退缩，以及随着永久冻土活性层的厚度增加，永久冻土覆盖面积的缩小。从1693年开始编纂整理的芬兰托尔尼奥河的冰盖记录显示，现在冰冻河的春融时间提前了一个月。第五，我们跟踪天气记录，看到了显著的变化。近年来，大规模的风暴和随后的洪水袭击了中国、意大利、英国、韩国、孟加拉国、委内瑞拉、巴基斯坦、澳大利亚和莫桑比克。2011年，加拿大的闵胜基博士及其同事在《自然》期刊上发表的 111 对北半球所有降水记录的详细汇编支持这些观察结果，并显示过去60年中降雨强度显著增加。此外，在英国，2000年至2001

年冬季是18世纪开始记录以来最湿润的六个月；2008年8月是有记录以来最湿润的8月；2012年4月至6月是有记录以来最湿润的春季。同时英国公众收集到的数据显示，鸟类筑巢时间比35年前提前了12±4天。第六，我们分析了自然变化对气候的影响，包括太阳光斑和火山爆发，虽然这些对理解过去150年的温度变化模式至关重要，但并不能解释整体的变暖趋势。第七，我们现在了解了过去较长时期的气候变化，以及温室气体在决定地球气候方面所起的作用。

"气候门"

尽管有各种证据，但对未来气候变化的讨论还是引起了强烈的反应。一部分原因是，我们为改善这种状况而可能必须做出的许多改变似乎与西方当前的新自由主义市场驱动方式背道而驰。另一部分原因则是媒体、公众和政治家对科学的基本误解。马克·亨德森的《极客宣言》一书对此有很好的讨论。"气候门"和媒体报道的其他所谓的气候变化密谋就是这种误解的最好例子。因为科学不是一个信仰体系，人们不会（因为抗生素可能会拯救人命而）决定自己相信抗生素，也不相信乘坐有尖锐突起的金属管便可安全地飞越大西洋，但他们会否认吸烟导致癌症，或人类免疫缺陷病毒导致艾滋病，抑或是温室气体导致全球变暖。科学是一种基于收集和积累证据、能够自我修正的理性方法论，这是我们社会的基础。2009年11月，在"气候门"事件中，东英吉利大学（UEA）气候研究室（CRU）的数千封电子邮件和其他文件因黑客攻击而被非法公布。据指控，这些电子邮件揭示了气候科学界的不当行为，包括隐瞒科学信息、阻止

112

论文发表、删除原始数据,以及操纵数据等,让全球变暖的论调看起来比实际情况更有力。三项独立调查的结论是,没有证据表明存在科学渎职行为。但当时所有媒体的评论员都忽略了一点,即NOAA与NASA的另外两个主要小组使用了不同的原始数据集和不同的统计方法,却发表了与UEA小组完全相同的结论。这一点在2012年得到了进一步的证实,曾是气候变化怀疑论者的物理学家理查德·穆勒教授和他在加州大学伯克利分校的小组在那一年公布了他们整理的过去200年的全球温度记录,他公开宣布自己改变了想法,认为气候变化**的确是**由于人类活动而发生的。

图37显示了过去2 000年来全球温度的所有这些综合数据集,毫不奇怪,它们各不相同,但显示出非常相似的趋势,都表明20世纪比过去2 000年的其他任何时间都要温暖。

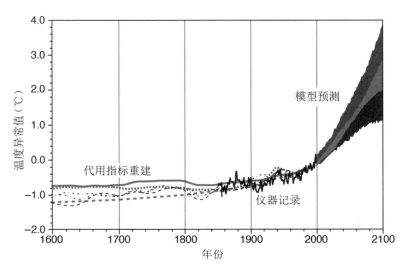

图37 过去和未来的全球平均表面温度

还有人指责UEA小组和其他气候科学家篡改了原始数据。科学家们使用的简短术语,如"纠正""欺骗""调整""操纵""一条线"和"相关"等,当然对这一点帮助不大。然而,一些原始数据确实需要处理,以便与其他数据进行比较,特别是在试图对温度进行长时间的记录,而在此期间用于测量温度的方法已经改变的时候。最明显的例子就是测量海温,在1941年之前,都是通过测量吊在甲板上的水桶里的海水来获得海温的。最初使用的都是木制水桶,随后在1856年至1910年期间改用帆布桶。因为水桶被吊在甲板上,这种设备的变化会影响水由于蒸发而造成的冷却程度。此外,这一时期经历了从帆船向蒸汽船的逐步转变,改变了船甲板的高度和船速,两者均会影响水的蒸发冷却。自1941年以来,大多数海温测量都是在船舶的发动机进水口进行的——同样,这又是一次转变。如果科学家只是把这些原始数据都归在一处,当然会出错。此外在这种情况下,因为早期的海面温度测量值太低,如果不加以修正,就会使海洋的全球变暖显得比实际情况严重得多。所以在科学的各个环节中,对数据的不断检查和修正极其重要。但最重要的部分是结果是否可以重现:是否有来自许多研究小组的重量级证据显示出变化?这就是在对气候变化进行了30多年的深入研究后,大多数科学家对气候变化正在发生,而且是由人类活动造成的这一结论有很大把握的原因。

气候变化及其影响

在第三章中,我们探讨了科学家如何模拟气候和未来气候的变化。IPCC的科学报告中介绍了所有环流模型的综合情况。

他们在2007年报告说，到2100年，全球平均地表温度可能上升1.1 ～ 6.4 ℃，最准确的估计是1.8 ～ 4.0 ℃。影响气候变暖的最大因素是排放的影响，以及采用何种场景。温室气体排放量上升得越快越高，世界就会变得越热。需要注意的是，尽管全球经济衰退，但全球二氧化碳排放量的上升速度，与已经探讨过、IPCC最可怕的"一切照旧"排放场景一样快。该模型还预测，全球平均海平面将上升18 ～ 59厘米。如果将格陵兰岛和南极洲融冰的贡献计算在内，那么到2100年，这一范围将增加到28 ～ 79厘米。所有这些预测都假设全球气温和冰盖损耗之间存在持续的线性反应。但这不可能，因此海平面上升的幅度可能会大得多。IPCC的下一份科学报告将于2013年发布，并采用更现实的未来排放路径，但该报告的初稿章节显示，他们会得出与2007年的报告非常相似的结论。

随着地球温度的上升，气候变化的影响将显著增强。洪水、干旱、热浪和风暴的重现期与严重性也将增加。随着海平面的上升，沿海市镇会特别脆弱，加剧了洪水和风暴潮的影响。2009年《柳叶刀》期刊发表的伦敦大学学院最近的一项多学科研究表明，气候变化对人类健康的最大威胁来自水和粮食安全性的降低，这可能会影响数十亿人。气候变化还威胁着世界上已经遭到破坏的生物多样性。生态系统已经因栖息地的丧失、城市化、污染和狩猎而严重退化。2007年的《千年生态系统评估》报告显示，每小时就有三种已知生物灭绝，而2008年的《地球生命力指数》显示，全球脊椎动物的生物多样性在短短35年内减少了三分之一以上，这一灭绝速度比化石记录中观察到的任何物种都要快10 000倍。英国皇家学会在2012年发表了一份精彩的《人与地球》报

115

告,总结了人类对环境造成的巨大影响,随着全球人口的增加,更重要的是随着全球消费不受控制地持续增长,这种情况将变得更加严重。当然,气候变化将加剧所有这些环境退化。

气候变化的"安全"水平是什么?

那么,什么程度的气候变化才算"安全"呢? 2005年2月,英国政府在埃克塞特召开了一次国际科学会议讨论这个话题。这样一来,当时的首相托尼·布莱尔就有了一个政治性的数字,可以在当年晚些时候由英国主办的八国集团首脑会议上提出。他们的建议是,全球变暖必须限制在比工业化前的平均温度高2 ℃的范围内。低于这个临界点,由于区域性的气候变化,似乎既有赢家也有输家,但高于这个数字的话,看来人人都会输。但这只是一个纯粹的政治视角,因为如果你住在任何一座低洼的太平洋岛屿上,那么当达到这个2 ℃的水平时,整座岛屿可能已经被淹没了。然而,随着新的气候条约的失败,如今看来,气温的上升甚至有可能超过这个临界点。目前,在"一切照旧"的排放场景下,我们将在远没有到2050年时就达到2 ℃的临界点。这并不奇怪,因为国际能源署预测,未来20年化石燃料的使用将包括石油用量增加30%,煤炭用量增加50%,天然气用量增加40%。

那么拯救世界的代价是什么呢? 根据2006年英国政府委托的"气候变化经济学"《斯特恩报告》,如果现在尽一切可能减少全球温室气体排放,并确保我们适应气候变化带来的迫在眉睫的影响,那么每年只需花费世界国内生产总值的1%。然而,如果我们无动于衷,那么气候变化的影响可能会使我们每年损

116

失世界国内生产总值的5%～20%。这些数字一直存在争议。一些专家认为，由于全球排放量的增长速度比最坏的预测还要快，因此将全球经济转变为低碳排放的成本可能超过世界国内生产总值的1%。对此，尼古拉斯·斯特恩爵士最近将他的数字修订为世界国内生产总值的2%。另一些人则认为，这些成本可以通过区域碳交易系统来抵消。还有人认为，IPCC和《斯特恩报告》低估了全球变暖的影响和相关成本。然而，即使解决全球变暖的成本效益低于《斯特恩报告》的建议，也有明显的道德理由来防止数千万人死亡，以及未能解决全球变暖而造成数十亿人的苦难加剧。

尽管达成全球气候变化新条约的努力遭到了彻底的失败，但还是有一些国家和地区在认真对待这些报告。英国出台了具有长期法律约束力的《气候变化法案》。该法案提供了一个法律框架，确保政府实现到2050年温室气体排放量比1990年的水平至少减少80%的目标。2012年春，墨西哥也加入了英国的行列，批准了本国的气候变化国家法律，明确到2050年将减少50%的排放量。在欧盟，所有的成员国都同意到2020年实行"20：20：20"政策。这是为了实现温室气体排放量减少20%，能源效率提高20%，以及所有能源的20%由可再生资源生产这一目标而做出的努力。

117

总　结

到2030年，全球粮食和能源需求将增加50%，水需求将增加30%。部分原因是全球人口的增加，但主要是由于低收入国家的快速发展造成的。此外，气候变化的影响越来越大，直接威胁到水和粮食的安全,于是就有了约翰·贝丁顿爵士(英国政府首席科学顾问)所说的"完美风暴"(图38)。因此,气候变化和可持续能源是21世纪的关键科学问题。我们已经有明确的证据表明,在20世纪,全球气温上升了0.75 ℃,海平面上升了22厘米。IPCC预测,到2100年全球气温可能上升1.8 ~ 4.0 ℃,这个范围取决于我们在未来90年内排放多少温室气体的不确定性。如果格陵兰冰盖和南极冰盖加速融化,海平面可能会上升得更多,达到28 ~ 79厘米。此外,天气模式将变得更难预测,风暴、洪水、热浪和干旱等极端气候事件的发生将变得更加频繁。在下一章中,我们将看一看修复气候的所有不同选择。

图38　完美风暴:预计到2030年需求量的增加

第九章

应对气候变化

引　言

　　为了防止未来气候变化最坏的影响，最明智的办法是减少温室气体排放。科学家认为，为避免最坏的影响，到21世纪中叶，二氧化碳的排放量必须减少50%～80%。然而许多人认为，大幅削减化石燃料的使用，将严重影响全球经济。在一个每年有800万儿童无谓地死亡、8亿人每晚饿着肚子睡觉、10亿人仍然无法经常获得清洁安全的饮用水的世界上，这将阻碍快速发展以及全球贫困的减少。尽管如此，在"一切照旧"的排放场景中，我们很有可能在2100年之前面临至少4℃的升温，这不啻是一场灾难，其影响将不成比例地落在社会中最贫穷的人身上。在本章中，我们将探讨解决气候问题的三个主要方法。第一是"减缓"或降低我们的碳排放量；第二是从源头或从大气中清除二氧化碳；第三是利用技术减少地球吸收的太阳辐射，从而使地球降温。

120

减　缓

在未来35年内将全球碳排放量至少减少一半，并在21世纪末将碳排放量减少80%，这听起来像是幻想，特别是考虑到目前化石燃料使用的趋势（图39）。然而，普林斯顿大学的史蒂夫·帕卡拉和罗伯特·索科洛在《科学》期刊上发表了一篇很有影响力的论文，试图让这一挑战看起来更容易实现。他们将"一切照旧"的排放场景与理想的450 ppmv的方案进行了对比，并将两者之间的差异描述为一些"楔子"。与其说我们看到的是一个不可克服的巨大问题，不如说我们真正要实现的是许多中等规模的变化，这些变化累加起来就是巨变。他们提供了几个楔子的实例，每个楔子每年可封存约10亿兆吨的碳。例如，一个楔子就是将20亿辆汽车的效率提高一倍，从每加仑30英里（30 mpg）提高到60 mpg，这其实是一个很容易实现的目标，因为已经有一些家用汽车可以轻松达到100 mpg了。

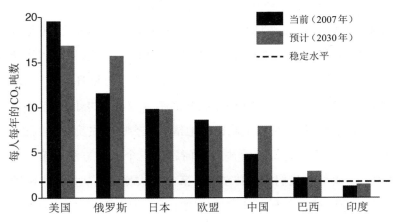

图39　按国家分列的过去和未来的碳排放量

一个楔子就是提高能源效率。目前,美国一个普通家庭的能源用量是加利福尼亚州普通家庭的两倍,而加州家庭能源的用量又是丹麦普通家庭的两倍。因此,发达国家只要提高能源效率,就可以节省大量的能源。如果工业和商业部门能够减少能源的使用,就可以大大降低运营成本;但由于目前能源成本非常低,这还不是他们的当务之急。无论如何,遗憾的是这些部门取得的一切效率提升最终都可能被增产所利用,能源的用量也会相应提高。例如,如果我们确实将20亿辆汽车的效率提高了一倍,那么如果再增加20亿辆汽车的产量,这一收益就会被抹去。所以有人认为,减少二氧化碳的生产,重中之重应该是生产清洁能源或无碳能源。这一点将在下一节讨论。

替代能源

利用化石燃料提供能源是一个惊人的发现,使我们的发展速度超过史上的任何时期。发达世界的高生活水平正是建立在使用这些廉价且相对安全的化石燃料的基础上。遗憾的是,我们发现,燃烧化石燃料会产生意想不到的后果,即改变全球的气候。这是因为燃烧这些化石燃料时,我们在释放数百万年前被锁住的更多的太阳能。相反,我们需要从使用过去气候条件下产生的能源转为使用当前气候系统的能源。这些能源包括太阳能、风能、水能、波浪能和潮汐能。我们改用可再生能源还有其他的原因。首先,有人担心我们的石油供应已经触顶,世界上的这种化石燃料行将耗尽。煤可能也是一样,尽管还剩下可用几百年的高品位煤。其次,21世纪各国对"能源安

122

全"的认识已经非常深刻：大多数发达国家的经济严重依赖进口化石燃料，这使得它们非常容易受到市场波动和国际讹诈的影响。

下面将简要讨论主要的替代能源。显而易见，某些替代能源适合相应的国家。例如，英国拥有整个欧洲最好的风力资源，沙特阿拉伯则拥有获取太阳能的绝佳条件。几乎所有的替代能源都利用了气候系统。

太阳能

在第一章中，我们讨论了地球从太阳获得平均每平方米343瓦特（343 W/m²）的能量，而整个地球只接收了太阳发出的所有能量的20亿分之一。所以太阳在很多方面都是植物已经利用了数十亿年的能源终极来源。目前我们可以将太阳能直接转化为热能或电能，也可以通过种植生物燃料，利用光合作用获取能量。最简单的方法是"太阳能供暖"。小规模的做法是，例如在阳光充足的国家，通过在房屋和其他建筑的顶上安装太阳能加热板把水加热，人们就可以享受无碳的热水淋浴和盆浴。大规模的做法可以利用抛物柱面镜来聚焦太阳能，产生热的液体（水或油），从而驱动涡轮机发电。太阳能热电厂的最佳选址是低纬度沙漠，那里每年的阴天很少。自1980年代以来，太阳能热电厂已在加利福尼亚州建成并使用，现在其他许多国家也在建设和使用太阳能热电厂。太阳能光电板或太阳能电池板将阳光直接转化为电能。一束束阳光照射到太阳能电池板上，驱动其内部的电子，产生电流。太阳能电池板的主要优点是可以将其放置在需要能源的地方，避免了通常需要

123

的复杂基础设施来输送电力。过去十年，它们的效率有了很大的提高，最好的商用太阳能电池板的效率约为21%，大大高于光合作用约为1%的效率。随着对该技术的巨大投入，太阳能电池板的价格有了显著下降。

生物燃料

生物燃料是通过光合作用将太阳能转化为植物生物质而产生的，接下来可用于生产液体燃料。全球经济的基础是使用液体化石燃料，交通运输部门尤其如此。所以，从植物中提取的燃料在短期内可以成为发动汽车、船舶和飞机的中级低碳方式。问题在于生物燃料的生产可能会与粮食作物的生产竞争。事实上，2008年和2011年时粮食价格高点的出现最初就被归咎于生物燃料的生产。然而，新英格兰复杂性研究所的分析表明，超过50%的价格大幅上涨，其原因实际上是金融市场对粮食价格的投机。

最后，电动汽车是未来的发展趋势，因为我们可以保证生产的电力是碳中和的。然而对于飞机来说，这不是一个选项。传统的飞机燃料是煤油，它兼具了相对较轻的质量和较高的能量输出的优点。相关研究目前正在进行，以确定是否可以生产一种质量和能量足以取代煤油的生物燃料。

风　能

风力涡轮机是一种有效的发电手段——如果它们足够大并且最好位于海上的话。理想情况下，我们希望涡轮机的大小与自由女神像一样，以达到最大的效果。例如，泰晤士河口正在建造"伦敦阵列",将产生1 000兆瓦（1 000 MW）的电力，使其成为获批的世界上最大的海上风电场。建成后，它可以为75万户家庭供电——约占大伦敦地区的四分之一——每年可减少140万吨的有害二氧化碳排放。

风力涡轮机存在一些问题。首先，它们无法提供持续的电力供应：不刮风就没有电。其次，人们不喜欢它们，认为它们丑陋、噪声大，还有人担心对当地自然栖息地产生影响。所有这些问题都很容易克服，只要把风电厂设在偏远的地方，比如外海，远离有特殊科学或自然价值的地区。一项研究表明，风力发电原则上可以产生逾12.5万太瓦时电量，是目前全球电力需求的五倍。

波浪能和潮汐能

波浪能和潮汐能或许是未来的重要能源。它们的概念很简单，就是将海洋以波浪形式的连续运动转化为电能。然而，这说来容易做来难，该领域的专家认为，现在的波浪能技术是太阳能电池板技术20年前的水平——需要迎头赶上。但与太阳能和风能相比，潮汐发电尤其有一个关键优势：它是恒定的。在任何一个国家，为了维持供能，必须保证至少20%的发电量，这被称为"基线"。随着向替代能源的转变，需要开发新的电力来源，以确保这一恒定的基线水平。

水 能

水力发电是全球重要的能源。2010年，水电供应了世界能源的5%。大部分电力来自大型水坝项目。这些项目可能会带

来重大的道德问题，因为大坝上方的大片土地必然被淹，导致大
规模人口迁移，破坏当地环境。大坝还会减慢河水的流速，影响
营养丰富的淤泥在河道下游地区的沉积。如果河流跨越国界，
就有可能出现水权和淤泥权的问题。例如，孟加拉国下沉的原
因之一是淤泥匮乏，而这种匮乏是由印度主要河流上的水坝造
成的。此外，还有一个关于水电站能减排多少温室气体的争论，
因为尽管发电不会造成任何碳排放，但大坝后面被淹地区的腐
烂植被仍会释放出大量的甲烷。

　　还有其他不依赖气候的替代能源或低碳能源，为了完整起
见，下一节将简要讨论这些能源。

地热能

　　在我们脚下的地球深处是炙热熔融的岩石。在诸如冰岛
和肯尼亚等一些地方，这种高温岩石非常接近地球表面，可以用
来加热水以产生蒸汽。这是一种极好的无碳能源，因为蒸汽产
生的电可以用来把水抽到高温岩石上。遗憾的是，这种能源受
到地理环境的限制。不过，还有另一种方式可以利用这种地热。
所有的新建筑都可以在其下方设置一个带有地源热泵的钻孔。
在这个系统中，冷水被抽进这些井眼里，下方的炙热熔岩就会把
水加热，从而降低加热水的成本，这种技术几乎可以应用于世界
任何地方。

核裂变

　　铀等重原子分裂时会产生能量，这就是所谓的核裂变。核
裂变的直接碳排放量很低，但铀的开采和后期停用的电站都会

产生大量的碳。目前，全球5%的能源来自核电。新一代核电站的效率极高，达到接近90%的水平。核电的主要缺点是会产生高放射性废物和安全问题。然而，效率的提高减少了废物，新一代核反应堆的所在地都有最先进的安全防范措施。核电的优点在于可靠，可以为能源组合提供所需的基线；而且核电是一种随时可用的技术，并已经经过了全面的测试。

核聚变

核聚变指两个较小的原子融合在一起时产生能量。太阳和其他恒星的能量都是这样来的。这个理念是指在海水中发现的重形式的氢可以结合起来，唯一的废物是非放射性的气体氦。当然，问题是如何让这两个原子结合在一起。太阳通过将原子置于难以置信的高温高压之下实现了这一点。英国的欧洲联合环状反应堆（JET）项目已取得了一些进展，该项目已经产生了16兆瓦的核聚变电力。问题在于最初产生高温所需能量的多少，以及将生产规模扩大到发电厂水平的难度。

清除二氧化碳

"地球工程"是指可用于清除大气中的温室气体或改变地球气候的技术的总称（图40）。被归于地球工程项目下的理念五花八门，既有非常明智的，也有彻底疯狂的。目前，我们每年释放超过85亿吨的碳（8.5 GtC），因此任何干预措施都必须能够在很大的范围内进行。在本节中，我们将研究如何清除和处理大气中的二氧化碳。主要的方法有三种：生物手段、物理手段和化学手段。

127

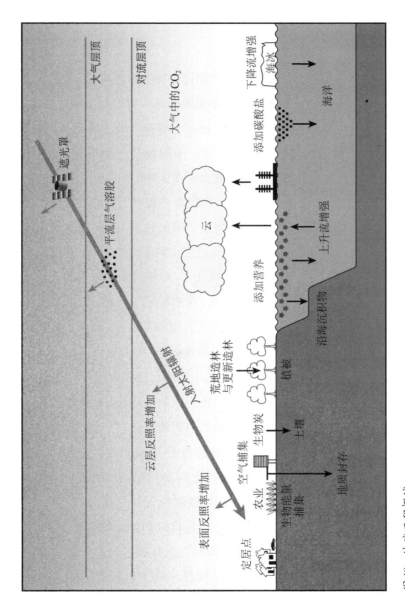

图 40 地球工程概述

125

生物除碳

除碳的生物方法包括使用前面讨论的生物燃料，以及更新造林。更新造林或荒地造林以及避免毁林是明智的双赢解决方案。维护森林可以保持生物多样性，稳定土壤和当地的降雨量，并通过碳信用额度为当地人提供生计。中国就是一个很好的例子。到1990年，黄土高原（此地在过去至少3 000年间曾是中国的粮仓）正在变成一个干旱尘暴区。砍伐森林和对土壤的过度耕作使其肥力开始下降，于是农民砍伐了更多的树木，开辟更多的土地来生产足敷生存的粮食。中国政府意识到了这个问题，并在随后的10年里推行了一项有力的植树计划，对任何砍伐树木的人进行严厉的惩罚。这些措施效果惊人，树木稳定了土壤，大大减少了土壤侵蚀。树木通过蒸腾作用为大气增加了水分，减少了蒸发和水分流失。一旦森林达到临界规模和面积，降雨量也开始稳定下来。在第八章中，我们看到土地的生物圈已在吸收每年约20亿吨的污染，史蒂夫·帕卡拉和罗伯特·索科洛估计，如果我们完全停止在全球砍伐森林，并将目前的植树速度提高一倍，就可以产生他们所说的另一个每年10亿吨的楔子，并获得更新造林的所有双赢利益。在英国，林业委员会提出到2050年将林地从12%增加到16%。这就意味着，由于森林对碳的吸收和封存，政府到2050年将二氧化碳排放量减少80%的目标将只剩70%就能实现了。

129　　　第二种生物方法是改变海洋对二氧化碳的吸收。最著名的海洋"技术性处理"是由已故的约翰·马丁教授提出的。他认为世界上许多海洋的生产力不足，因为海洋中缺乏重要的微量养分，其中最重要的是铁，它能使植物在表层水域中生长。海洋植物需要微量的铁，否则就无法生长。在大多数海洋中，有足够的富含铁的尘埃从陆上吹来，但太平洋和南大洋的大片区域似乎并没有收到多少尘埃，因此铁元素贫乏。所以有人建议用铁给海洋施肥，刺激海洋生产力。额外的光合作用会将更多的地表水二氧化碳转化为有机物。此外，当生物死亡时，有机物会沉降到海洋底部，带走并封存额外的碳。减少的表层水二氧化碳就会被大气中的二氧化碳补充。所以简而言之，给全世界的海洋施肥有助于清除大气中的二氧化碳，并将其封存在深海沉积物中。海上的实验结果差异很大，有的实验显示完全没有效果，而有的则显示需要大量的铁。此外，一旦停止添加额外的铁，这些封存的二氧化碳大多会被释放出来，因为每年只有极少量的有机物能够从透光带（海洋中足够接近海面的部分，可以接收光合作用所需的阳光）中逸出。

物理除碳

在工业生产过程中清除二氧化碳是一件棘手而又昂贵的事情，因为不仅需要清除二氧化碳，而且必须将其安全封存。每吨二氧化碳的清除和封存成本可能在 10～50 美元之间。这将导致电力生产成本增加 15%～100%。然而，由于廉价丰富的高品位煤供应，碳捕集与封存（CCS）被认为是世界各国政府最大的希望之一。为了使 CCS 高效运作，并使其达到一个可负担的水平，还需要在这个领域开展更多的研究。但问题是，非 CCS 的煤炭或天然气生产的电力始终比 CCS 生产的便宜，因此需要立法确保企业有义务使用 CCS 或转向替代能源生产。例如，欧盟的碳排放交易体系（ETS）是一个"限额与交易"系统，所有生产

130

或使用大量能源的公司都必须参与其中，以助力欧盟到2020年减少总排放量。

　　另一个可能的解决方案是直接从空气中清除二氧化碳。尽管二氧化碳仅占大气的0.04%，但此事说来容易做来难。一个疯狂的想法是生产人造树或塑料树。理论物理学家克劳斯·拉克纳和工程师艾伦·赖特在气候学家沃利·布勒克尔的支持下，设计出了可与二氧化碳结合的塑料，能将二氧化碳从大气中清除出去。在拟议的系统中，二氧化碳随后会从塑料中释放出来，并被带走封存。第一个问题是水，因为塑料受潮后会将二氧化碳释放到溶液中。这意味着塑料树必须放置在非常干旱的地区，或者需要巨大的雨伞来保护它们免遭雨淋。第二个问题是建造、运行、封存二氧化碳所需的能源量。第三个问题是规模问题：仅仅是为了应对美国的碳排放，就需要数以千万计的巨大人造树。在电台节目里就这个方法展开辩论时，我非常温和地建议，也许我们种植普通的树木就可以了。

　　然而，如果塑料树不是解决办法，就可能需要另一种形式的技术，从源头或最终从大气中清除二氧化碳。

风化除碳

131　　二氧化碳通过千百年的风化过程才能从大气中自然去除。第六章探讨了这一过程，当时我们讨论了喜马拉雅山脉的隆起对于清除大气中二氧化碳的作用。大气中的二氧化碳（CO_2）可以直接与硅酸钙（$CaSiO_3$）发生反应：

$$CaSiO_3 + CO_2 \longrightarrow CaCO_3 + SiO_2$$

这是一个极其缓慢的过程，每年去除不到1亿吨的碳，比我们的排放量少一百倍。另一个过程是利用雨水和二氧化碳的结合，形成弱碳酸溶液：

$$CaSiO_3 + 2CO_2 + H_2O \longrightarrow Ca^{2+} + 2HCO_3^- + SiO_2$$

只有硅酸盐矿物的风化作用才会对大气中的二氧化碳含量产生影响，因为碳酸盐岩石的风化作用会使二氧化碳返回大气。影响硅酸盐矿物的水解反应的副产品是生物碳酸盐（HCO_3^-），经海洋浮游生物代谢后转化为碳酸钙。海洋生物群的方解石质残骸最终以深海沉积物的形式沉积下来，从而在其所处的海洋地壳生命周期内从全球生物地球化学碳循环中消失。

有一些旨在加强这些自然风化反应的地球工程理念。一项建议是在用于农业的土壤中添加硅酸盐矿物。这将清除大气中的二氧化碳，并将其固定为碳酸盐矿物和生物碳酸盐溶液。然而，这一做法规模巨大，对土壤及其肥力的影响尚不清楚。另一个建议是提高二氧化碳与地壳中玄武岩和橄榄石的反应速度。浓缩的二氧化碳将被注入地下，并在地下深处形成碳酸盐。同样，像许多地球工程的理念一样，这是一个很好的建议，但研究它是否可行、安全，以及在必要的规模上进行的可能性工作仍少之又少。

132

碳封存

并非所有回收的二氧化碳都必须封存；有些可用于提高石油采收率，促进食品工业、化学制造（生产纯碱、尿素和甲醇）和金属加工工业发展。二氧化碳还可以应用于建筑材料、溶剂、清洁化合物、包装,以及废水的处理。但实际上,从工业过程中捕集的大部分二氧化碳都必须封存起来。据估计,从理论上讲,世界石油和天然气总储量燃烧形成的二氧化碳有三分之二可以封存在相应的储层中。其他的推测表明,仅天然气田就可封存900亿~4 000亿吨,含水层中还可封存900亿吨。

海洋也可用于处理二氧化碳。已提出的建议包括通过水合物倾倒进行封存——如果将二氧化碳与水在高压和低温下混合,就会产生一种固体或水合物,它比周围的水更重,因此会沉到海底。近来的另一个建议是将二氧化碳注入半英里深的巨大熔岩流之间的破碎火山岩中。二氧化碳将与渗透到岩石中的水发生反应。酸化的水会溶解岩石中的金属,其中主要是钙和铝。一旦它与钙形成碳酸氢钙,就无法再冒出水面逃逸了。纵使它会逸入海洋,碳酸氢钙也是相对无害的。对于海洋封存,还有一个我们在第二章中看到的复杂情况:海洋是循环的,所以无论你倾倒了多少二氧化碳,最终都会有一部分返回大气。此外,关于这将对海洋生态系统产生何种环境影响,科学家们还很不确定。

所有这些封存方法的主要问题都是安全问题。二氧化碳是一种非常危险的气体,因为它比空气重,可以造成窒息。一个重要的例子是1986年发生在喀麦隆西部尼奥斯湖的二氧化碳大爆炸,造成了逾1 700人和大量牲畜的死亡,波及范围远至25千米之外。虽然以前也发生过类似的灾难,但从来没有在一次短暂的事件中造成这么多的人和动物如此大规模地窒息。科学家们现在认为,当时发生的情况是,附近火山中溶解的二氧化碳从湖下的泉水中渗出,在上方水的重压之下被困在了深水中。1986年,一场雪崩把湖水混杂在一起,导致整个湖区的爆炸性翻涌，被困的二氧化碳一下子全部释放了出来。然而，美国各地不断有大量矿藏的古代二氧化碳被抽出，以提高石油采收率。目前还没有任何重大事故的报告，维护这些管道的工程师们认为，这些管道比贯穿大多数主要城市的天然气或石油管道要安全得多。

太阳辐射管理

正如我们看到的那样，拟议的许多地球工程解决方案仍然只是一些想法，需要做更多的工作来确定它们是否可行。已经提出的太阳辐射管理设想更是如此，其中许多想法听起来就像糟糕的好莱坞B级片里的东西。这些建议包括改变地球的反照率：增加反射回太空的太阳能以平衡全球变暖产生的热量（图40）。实现这一目标的方法包括：在太空竖起巨大的镜子；向大气层注入气溶胶；使农作物更具反射性；把所有的屋顶涂成白色；增加白云覆盖率；用反光的聚乙烯铝板覆盖世界上的大片沙漠。所有这些方法的根本问题是，我们不知道它们会产生怎样的连锁反应。目前，我们正在进行有史以来规模最大的地球工程实验之一，向大气中注入大量的温室气体，虽然我们对可能发生的情况有所了解，但仍不知道这会对气候系统产生什么具体的影响。这些地球工程的解决方案也同样如此——我们目前对它们是否会奏效，或者可能会产生哪些始料未及的副作用，都近乎一无所知。在许多方面，气候变化对地球的影响可以比作疾病和人体：预防疾病总是比试图治疗疾病更可取，我们都知道药物和化疗或放疗的潜在副作用。

仅以其中一个比较牵强的想法为例，来说明太阳辐射管理的问题：在太空中使用镜子来偏转太阳光。其中最复杂的建议是亚利桑那大学天文自适应光学中心主任罗杰·安杰尔提出的，他建议使用由微小反射器所组成的网络，使部分光线偏离地球。他自己也承认，这一方法十分昂贵，需要16万亿艘微光飞船，成本至少1万亿美元，得花30年才能发射完毕。但就像我们讨论过的其他所有旨在改变地球反照率的想法一样，它不会奏效。原因在于，所有这些方法都囿于降低地球的平均温度，而忽略了当前温度分布的重要性，正如我们在第二章中所知道的，当前的温度分布是气候的驱动力。事实上，布里斯托尔大学的丹·伦特及其同事利用气候模型表明，这些方法会把我们带入一个完全不同的全球气候：热带地区将下降1.5 ℃；高纬度地区将上升1.5 ℃；与前工业时代相比，全球降水量会下降5%。

地球工程治理

英国皇家学会2009年关于地球工程的报告不仅审查了这一领域的现有科学材料，还迈出了重要的一步，试图了解与全球气候系统有关的治理问题。在考虑区域和全球气候的变化如何对各国产生不同影响时，存在大量的道德问题。虽然总体上的积极结果有可能出现，但降雨模式的细微变化或许意味着整个国家都得不到足够的雨水，或是降雨过量导致灾害。皇家学会总结了关于地球工程的三种主要观点：

1. 这是一条争取时间的路线，以便让失败的国际减排谈判迎头赶上；

2. 它是对地球系统的危险操纵，可能在本质上并不道德；

3. 严格来说，它是针对主要减缓政策失败的一项保险政策。

与许多新兴的现代技术领域一样，即使允许开展研究且必须实施地球工程解决方案，也需要新的灵活治理和监管框架。目前，许多国际条约都与地球工程有关，但似乎没有一种手段足以应用。因此，"修复"气候对我们的民族国家的世界观提出了挑战，未来将需要新的治理方式。

适 应

即使我们决定大幅减少碳排放，并尝试了所有可用的地球工程方案，气候仍然会出现一些变化。这是因为气温已经上升了0.76 ℃，就算我们能将大气中的二氧化碳浓度降低到2000年的水平，也至少会再增加0.6 ℃。考虑到国际谈判的失败和对地球工程严肃关注的不足，我们目前正走在"一切照旧"的道路上。这意味着在不久的将来，许多国家将受到气候变化的不利影响；在未来30年内，几乎所有国家都会受到影响。因此，如果我们不能解决气候问题，就意味着我们也必须制订计划以适应不断变化的新气候。每个国家的政府都需要研究本国环境和社会经济体系的脆弱性，并预测气候变化最有可能对本国造成的影响。

然而，气候变化的主要威胁在于它的不可预测性。人类可以在从沙漠到北极的几乎任何极端气候中生活，但只有当我们能够预测这种极端天气的范围时，才能生存下去。因此，适应其实就是每个国家或地区如何应对新的极端天气水平带来的新威胁。这种适应应该从现在就开始，因为基础设施的改变可能需要30多年，在民主国家尤其如此。例如，如果想通过建立更好的海岸防护堤或将某一特定地区的农田改回自然湿地，从而改变土地的用途，就可能需要长达20年的时间来研究和规划适当的措施；接着花10年时间完成全面的咨询和法律程序；再花10年时间实施这些改变；最后花10年时间进行自然恢复。目前保护伦敦免受洪水侵袭的泰晤士河大堤就是一个很好的例子。它是为了应对1953年的特大洪水而修建的，但直到31年后的1984年才正式启用。

另一个问题是，适应需要现在就投入资金；而许多国家根本没有这笔钱。由于大多数人都是活在当下，所以人们不愿意支付更多的税赋来保护未来的自己。尽管讨论的所有适应措施从长远来看都会为地方、国家和世界节省资金，但作为一个全球社会，我们仍然有一个非常短视的观点——通常以历届政府之间的几年时间来衡量。

每个政府可以立即做的一件事是建立对气候变化影响的评估。例如，英国有一个英国气候影响项目（http://www.ukcip.org.uk），该项目显示了未来100年气候变化可能给英国带来的影响。这个项目的对象是英国国家和地方政府、工业界、商业界、媒体和公众。如果每个政府都设立一个这样的项目，那么至少他们的公民就能获得信息，就其国家应如何适应气候变化做出明智的选择。

总　结

那么我们应该如何解决全球气候变化的问题呢？首先，有一个国际政治解决方案似乎是明智之举。目前，我们还没有达成2012年后的协议，而且正面临着全球碳排放的大幅增加（图41）。任何政治协议都必须包括保护发展中国家快速发展的措施。在道义上，最贫困国家的人民有权发展，并获得与发达国家相同的

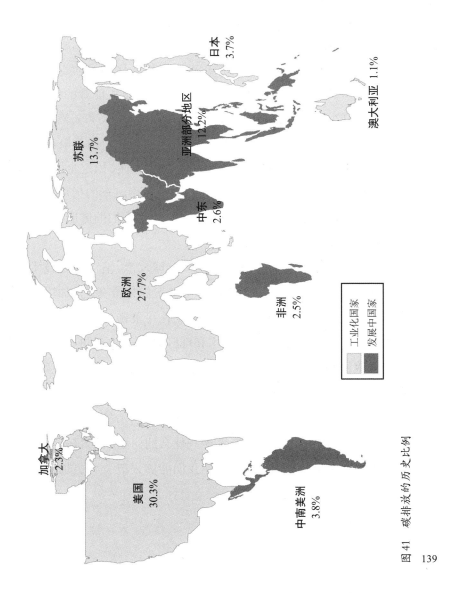

图 41 碳排放的历史比例

日本 3.7%

澳大利亚 1.1%

亚洲部分地区 12.2%

苏联 13.7%

中东 2.6%

欧洲 27.7%

非洲 2.5%

工业化国家
发展中国家

加拿大 2.3%

美国 30.3%

中南美洲 3.8%

139

135

生活方式。我们还需要对替代/可再生能源和低碳技术进行大规模投资，以提供减少世界碳排放的手段。我们应该投资地球工程解决方案，特别是那些将在短期内产生重大影响的解决方案，如更新造林和碳捕集与封存。应对气候变化的行动也应始终包含双赢的因素。例如，支持大幅增加对可再生能源的使用，不仅可以减少排放，而且可以减少对进口石油、煤炭和天然气的依赖，有助于保障能源安全。减少砍伐森林和更新造林既能减少大气中的二氧化碳，还有助于保持生物多样性、稳定土壤，并通过碳信用额度为当地人提供生计。减少使用汽车的措施将增加步行和骑自行车的人数，这反过来又会改善人们的健康，例如减少肥胖症和心脏病的发作。

气候 138

我们不能把所有的希望都寄托在全球政治、清洁能源技术和地球工程上——我们必须为最坏的情况做好准备，并加以适应。如果现在就实施，可以减轻气候变化可能造成的大部分成本和损失。但是，这需要各个国家和地区对未来50年进行规划，而由于政治制度的短视，大多数社会无法做到这一点。这意味着我们的气候问题正在挑战我们组织社会的方式。它们不仅质疑了民族国家与全球责任的概念，也暴露了政治领导人目光短浅。关于我们面对气候变化能够做些什么的问题，答案是，我们必须改变社会的一些基本规则，使我们能够采取一种更加全球性和长期可持续的方法。

140

终极气候变化

引　言

通过了解地球气候在过去是如何变化的，就有可能展望未来。本书中介绍的许多变化过程可以向前推演，看看近期和远期会发生什么。

下一个冰期

最新的科学研究告诉我们，新的冰期的缓慢降临会在未来1 000年内随时开始。正如在第七章中所看到的那样，我们了解了地球轨道的不同摆动是如何影响地球气候的。我们也可以回顾以前的间冰期，看看它们持续了多长时间。伦敦大学学院古气候学教授克罗尼斯·策达基斯及其同事计算了全新世时期的自然长度，并表明下一个冰期将会在从现在到未来1 500年内的任何时候开始。但他们也论证说这大概率不会发生。在其他的每一个间冰期，当气候系统走出冰期时，大气中的二氧化碳含量似乎都是最高的。大气中的二氧化碳含量在每个间冰期都会慢慢降低，直到一个约为240 ppmv的临界值，比前工业时代低40 ppmv，比现在低160 ppmv。一旦达到这个临界值，那么气候系统就可以对轨道驱动做出反应，并滑入下一个冰期。然而，大气中的二氧化碳污染如果保持在较高的水平，似乎可以阻止世界在未来1 500年内滑入冰期。事实上，根据比利时鲁汶天主教大学安德烈·贝尔热教授的模型预测，如果我们将大气中的二氧化碳浓度提高一倍，那么全球变暖会将下一个冰期再推迟4.5万年。然而有趣的是，到那时，轨道驱动将大到足以克服温室气体的极端水平，正常的冰期–间冰期旋回将重新确立。

另一个有趣的问题是，为什么在工业革命之前，大气中的二氧化碳含量就已经超过了我们预测的应该达到的水平？这就引出了拉迪曼奇妙的早期人类世假说。弗吉尼亚大学古气候学教授比尔·拉迪曼提出，早期的农艺师们从大约7 000年前开始逆转了大气中二氧化碳含量的自然下降，从大约5 000年前开始逆转了大气中甲烷含量的自然下降。这个论点引起了巨大的争议，但像所有的正确理论一样，它经过了一次又一次的检验，还没有人能够推翻它。所以本质上这个论点是，早期人类与环境的互动增加了大气中的温室气体，即使在工业革命之前，气候的变化也足以推迟下一个冰期的到来。这也提出了一个问题，即人类究竟是在什么时候成为这样一种地质力量的；由于所有的地质年代都有一个名称，一场运动由此发起，以定义人类开始对地球气候系统产生重大影响的年代。

已故的生态学家尤金·F. 施特默创造了"人类世"
（Anthropocene）一词，诺贝尔奖获得者、大气化学家保罗·克鲁岑将其推广。这个词源自希腊，anthropo意为"人类"，cene意为"新"。虽然这个词得到了很多支持，但我们仍然不知道全新世和人类世的界限在哪里。在地质学上，不同时期之间的界限必须有一个明确的基准点或"金道钉"，才能在世界范围得到公认。关于如何确定这个时期的年代，有人主张用比尔·拉迪曼提出的早期人类影响大气中二氧化碳的时期，有人主张用工业革命的时期，还有人主张利用冰芯中可以找到的文明的微量元素证据（例如，1960年代原子武器试验计划产生的氯层）来确定。无论最终选择何种基准，毫无疑问，人类现在是地球上的主要"地质"力量：通过包括砍伐森林在内的土地使用的大规模变化，改变了全球的侵蚀模式，造成了物种大量灭绝和生物多样性的巨大损失，改变了全球的氮循环、臭氧消耗，当然还有气候的变化。

下一个超级大陆

得克萨斯大学阿灵顿分校的克里斯托弗·斯科泰塞教授是"古代地图"项目的主任。该项目旨在说明过去10亿年来板块构造如何改变了海洋盆地和大陆及其位置。他们还推测了板块构造在未来将如何改变地球的面貌。在第五章中，我们看到了过去的超级大陆是如何形成的，以及它们对气候和演化的重大影响。根据"古代地图"项目，下一个超级大陆将在未来2.5亿年内形成。在未来的5 000万年之内，世界看起来与当前大致相同，但有些微变化：大西洋将继续变宽；非洲将撞上欧洲，地中海因而闭合；澳大利亚将撞上东南亚；加利福尼亚将沿着海岸线向北滑向阿拉斯加（图42）。真正有趣的变化发生在未来5 000万年到1.5亿年之间，其中的关键是南北美洲东海岸的重

143

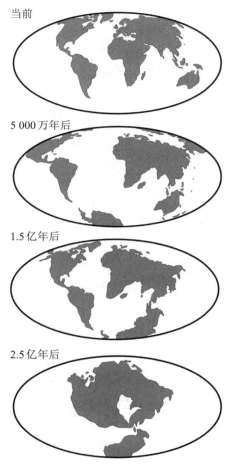

当前

5 000万年后

1.5亿年后

2.5亿年后

图42 未来各大洲的位置

大变化。目前这里是一个被动的陆缘，大陆和海洋板块是相连
的。但在未来5 000万年左右的某个时候，大西洋沿岸各大陆将
穿过加勒比海和科科斯板块边界，在大西洋形成一个新的俯冲
带（构造板块碰撞，将一个板块推到另一个板块之下）。这将形
成西大西洋海沟，大西洋大陆板块将开始被吞没。尽管大西洋

中脊继续产生新的海洋板块物质，但破坏最终将超过创造，大西洋将开始封闭。在未来约1.5亿年时，大西洋中脊将到达俯冲带并被吞没。由于没有新的洋壳产生，大西洋的封闭速度将会加快。在其他地方，英国和欧洲届时将能看到北太平洋；地中海山脉将达到最大高度；南极洲和澳大利亚将形成一个大洲。到了2.5亿年后的未来，南北美洲、非洲和亚洲将结合在一起形成一个超级大陆，并拥有一个和澳大利亚一样大的内陆海。一个小型的海洋通道将把这个超级大陆和南极洲-澳大利亚大陆分开。我们从古气候记录中知道，超级大陆对生命不利，比如96%的海洋物种和70%的陆生脊椎动物物种都在2.5亿年前的超级大陆二叠纪-三叠纪大灭绝事件中灭绝了。因此，再过2.5亿年，地球上存在的一切生命都将面临另一个重大的挑战期。

沸腾的海洋

我们认为，只要有大陆存在，超级大陆形成和分裂的周期就会持续下去。但在某一时刻，气候会恶化到让复杂的多细胞生物无法生存的程度。例如，自从太阳形成以来，它的能量输出一直在增加。太阳的光度每隔十亿年就会增加10%左右。詹姆斯·洛夫洛克教授提出，生命和气候系统之间的反馈已经改变了大气中的温室气体含量，以容纳过去几十亿年来的这种光度增加；这是他精彩的盖亚假说的核心论点。然而从地质学的角度来看，尽管我们在过去100年里尽了最大努力来扭转情况，温室气体含量仍然极低，所以地球的自我降温能力已经接近了极限。这意味着在未来的几十亿年里，地球的温度很可能会逐渐上升（图43）。当地球温度高到海洋开始蒸发，将大量的水分抽

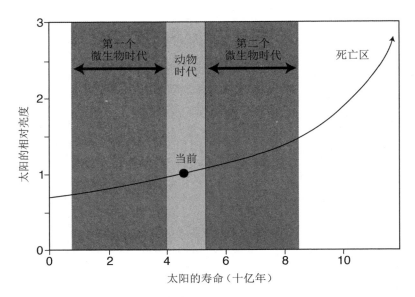

图 43 地球生命对比太阳的寿命

到大气中时,这样的一个临界点便会到来。正如我们在第二章中看到的那样,水蒸气是最重要的温室气体之一,这种失控的温室将使地球上的平均温度达到 100 ℃ 以上,目前所有的多细胞生物都无法生存。我认为,最终的气候变化将发生在海洋沸腾导致超级全球变暖时。生命以及复杂生命进化的地方,将成为这一切最终的毁灭者,这不啻是一种讽刺。就算最极端的微生物也只能再维持 30 亿年,到那时,即使对它们来说地球也太热了(图 43)。

地球之殇

虽然地球上的多细胞生命可能在大约 10 亿年后消亡,微生物也可能在大约 30 亿年后消亡,但这不会是地球的末日。据预

146

测，那一天将在大约50亿年后发生。太阳将其氢元素转化为氦元素的过程大约进行了一半。每一秒钟都有超过400万吨的物质在太阳的核心内转化为能量。在过去的45亿年里，太阳已将相当于100个地球质量的物质转化为能量了。然而，太阳并没有足够的质量来作为超新星爆炸。相反，在大约50亿年后，它理应进入红巨星阶段。核心的氢燃料消耗殆尽时，它的外层将会膨胀，核心收缩并升温。然后，氢聚变反应沿着围绕氦核的外壳继续进行，随着更多氦元素的产生，氦核稳步膨胀。一旦核心温度达到1亿摄氏度左右，太阳甚至会开始通过消耗氦元素来产生碳。作为红巨星，太阳将会比地球目前的轨道面还要大，大约是现在太阳半径的250倍。人们认为，即便如此，地球也有可能幸存下来，因为届时太阳的质量大约会减少30%，所以它的引力会减弱，导致周围行星的轨道向外移动。然而，墨西哥瓜纳华托大学的彼得·施罗德和英国萨塞克斯大学的罗伯特·史密斯创建了一个太阳向红巨星过渡的详细模型，他们发现，地球的轨道起初会变宽，但随后地球本身会因为自身的引力而在太阳表面造成一个"潮汐隆起"。这个隆起会在轨道上刚好滞后于地球，使其减速，足以将地球拖向燃烧毁灭。在红巨星阶段结束、地球消亡之后，太阳会抛掉外层形成行星状星云。抛掉外层后，唯一的残余之物就是极其炙热的恒星内核，再经过数十亿年的时间，恒星内核会慢慢冷却褪色，变成一颗白矮星。

147

索 引

（条目后的数字为原书页码，
见本书边码）

气候

H

I

J

K

L

气候

索引

气候

索引

气候

152

Mark Maslin

CLIMATE

A Very Short Introduction

Contents

Acknowledgements

The author would like to thank the following people: Anne, Chris, Johanna, Alexandra, Abbie Maslin, and Sue Andrews for being there; Emma Marchant and Latha Menon for their excellent editing and support; all the staff and friends at the UCL Environment Institute, UCL Department of Geography, TippingPoint, Rezatec Ltd, Permian, DMCii, KMatrix, and Global Precious Commodities; and last but not least Miles Irving for excellent illustrations.

List of illustrations

Climate

Chapter 1
What is climate?

Introduction

Climate affects everything we do in life, from the clothes we wear
to the diseases we catch. This is because as humans we only feel
comfortable within a very narrow range of temperature and
humidity. This comfort zone ranges from about 20°C to 26°C and
from 20 to 75 per cent relative humidity (see Figure 1). However,
we live almost everywhere in the world, meaning that conditions
are frequently outside this comfort zone, and we have learnt to
adapt our clothing and dwellings to maintain our comfort. So
while you may think the clothes you have hanging in your
wardrobe simply reflect your fashion taste or lack of, in reality
they reflect the climate in which you live and how it changes
throughout the year. So you have a thick padded coat for a
Canadian winter and a short-sleeved shirt for a business meeting
in Rio. Our wardrobes also give hints about where we like to take
our holidays. If you are a budding Polar explorer then there will
very warm Arctic clothes hanging up—if you love sunning yourself
on the beach, then there will be shorts or a bikini instead.

Our houses are also built with a clear understanding of local
climate. In England almost all houses have central heating as the
outside temperature is usually below 20°C, but few have air
conditioning as temperatures rarely exceed 26°C. On the other

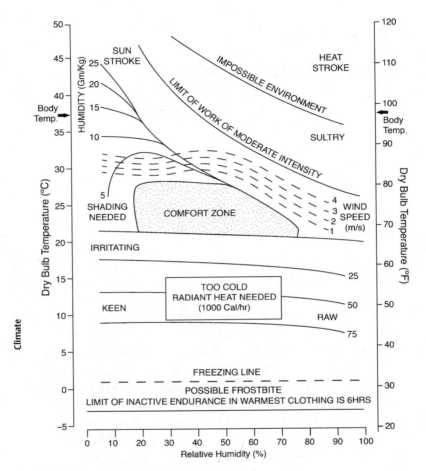

1. Human comfort and climate

hand, in Australia most houses have air conditioning but rarely central heating. Climate also affects the structure of our cities and how transport systems around the world operate. In Houston, Texas, there is a network of 7 miles of underground tunnels connecting all the major downtown buildings; this is fully climate controlled and links 95 heavily populated city blocks. People use the tunnel when it is raining or hot outside, because for at least 5 months of the year the average temperature in Houston is above

30 °C. Similarly there are underground malls in Canada to avoid the problems of heavy snow and extreme cold.

Climate controls where and when we get our food, because agriculture is controlled by rainfall, frost, and snow, and by how long the growing season is, which includes both the amount of sunlight and the length of the warm season. So in a simplified way, rice is grown where it is warm and very wet, while wheat can grow in much more temperate climes. The climate can also affect the quality of our food, for example it is well known that the very best vintages of French wine are produced when there are a few short sharp frosts during the winter, which harden the vines, producing excellent grapes. Farmers can also 'help' the local climate, for example by growing tomatoes in a greenhouse or by irrigating the land to provide a more constant supply of water.

Climate also influences where there will be extreme weather events such as heatwaves, droughts, floods, and storms. However in many cases our perception of extreme events is determined by local conditions, so for example in 2003 northern Europe was hit with a 'heatwave' and 100 °F (37.8 °C) was recorded for the first time ever in England. However in countries of the tropics a heatwave would not be recorded until temperatures were above 45 °C. Climate also has a large effect on our health, as many diseases are temperature and humidity controlled. For example incidences of influenza, commonly called the flu, reach a peak in winter. Since the Northern and Southern Hemispheres have winter at different times of the year, there are actually two different flu seasons globally each year. The influenza virus migrates between the two hemispheres after each winter, giving us time to produce new vaccinations based on the new strain of flu that has appeared in the previous six months in the other hemisphere. There have been many arguments about why flu is climate controlled and the theory is that during cold dry conditions the virus can survive on surfaces longer and so be more easily transmitted between people. Another suggestion is that vitamin D might provide some resistance or immunity to the virus. Hence in winter and during the tropical rainy season, when people stay indoors, away from the sun, their vitamin D levels fall and incidences of influenza increase.

Hot and cold Earth

The climate of our planet is caused by the Equator of the Earth receiving more of the sun's energy than the poles. If you imagine the Earth is a giant ball, the closest point to the sun is the middle or the Equator. The Equator is where the sun is most often directly overhead and it is here that the Earth receives the most energy. As you move further north or south away from the Equator, the surface of the Earth curves away from the sun, increasing the angle of the surface of the Earth relative to the sun. This means the sun's energy is spread over a larger area, and thus causes less warming. If we lived on a flat disc we would get much more energy from the

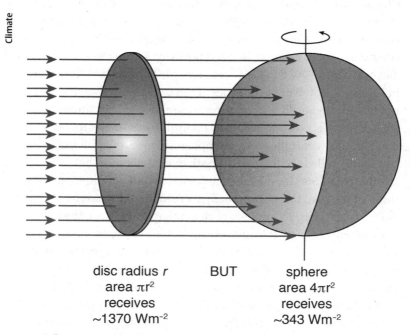

disc radius r BUT   sphere
area πr^2 area $4\pi r^2$
receives   receives
~1370 Wm^{-2} ~343 Wm^{-2}

2. **Solar energy distributed over a sphere**

sun—about 1,370 Watts per square metre (W/m²)—instead the planet surface averages about 343 W/m² (Figure 2) due to its curved nature. The Earth also receives a very small fraction of the energy pumped out of the sun. If you consider how small the Earth is compared with the sun, for every Watt we receive from the sun, it emits 2 billion Watts. This is why in many science fiction novels the authors imagine a strip or even a sphere around a star to collect all that energy that is simply being lost into space.

About one-third of the solar energy we receive is reflected straight back into space. This is because of 'albedo', which means how reflective is a surface. So, for example, white clouds and snow have a very high albedo and reflect almost all of the sunlight that falls on them, while darker surfaces such as the oceans, grassland, and rainforest absorb a lot more energy. Not only do the poles receive less energy than the Equator, but they also lose more energy back into space (Figure 3): the white snow and ice in the Arctic and Antarctic

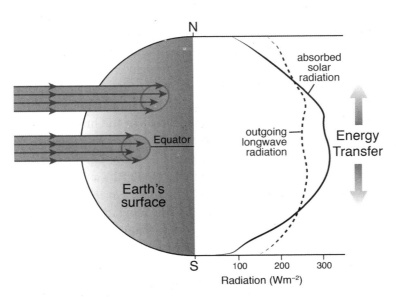

3. Energy transfer away from the Equator driven by angle of solar radiation

have a high albedo and bounce a lot of the sun energy back into space. On the other hand, the darker much less reflective vegetation at lower latitudes absorb a lot more energy. These two processes working together mean that the tropics are hot and the poles are very cold. Nature hates this sort of energy imbalance, so energy, in the form of heat, is transported by both the atmosphere and oceans from the Equator to both poles, and this affects the climate.

Earth in space

Our climate is controlled by two fundamental facts about the relationship between the Earth and the sun. The first is the tilt of the Earth's axis of rotation, which causes the seasons. The second factor is the daily rotation of the Earth that provides us with night and day and drives the circulation of both the atmosphere and the oceans.

The Earth's axis of rotation is tilted at an angle of $23.5°$ and results in a seasonal difference in the amount of energy received by each hemisphere throughout the year. The seasonal changes are by far the largest effect on climate. It is amazing to think that if the Earth were not tilted and stood straight up on its axis then we would not have spring, summer, autumn, and winter. We would not have the massive change in vegetation in the temperate latitudes and we would not have the monsoon and hurricane seasons in the tropics. The reason for the seasons is the change in the angle of the sunlight hitting the Earth through the year. If we take 21 December as an example, the Earth's axis is leaning away from the sun, so the sunlight hitting the Northern Hemisphere is at a greater angle and spreading its energy over a wider area. Moreover the lean is so great that in the Arctic the sunlight cannot even reach the surface and this produces 24 hours of darkness and winter in the Northern Hemisphere. However, everything is opposite in the Southern Hemisphere, since it is then leaning towards the sun and hence the sunlight is more directly overhead. This means that Antarctica is bathed in 24 hours of sunlight and people in Australia have Christmas dinner on the beach, while topping up their tan. As the Earth moves round the sun, taking about 365.25 days (hence the leap year every fourth year), the angle of the axis stays in the same place. Hence when it comes to June the Earth's axis is leaning towards the sun, so the Northern Hemisphere has lots of direct sunlight and thus summer, while the Southern Hemisphere is shielded from the sunlight and is plunged into winter.

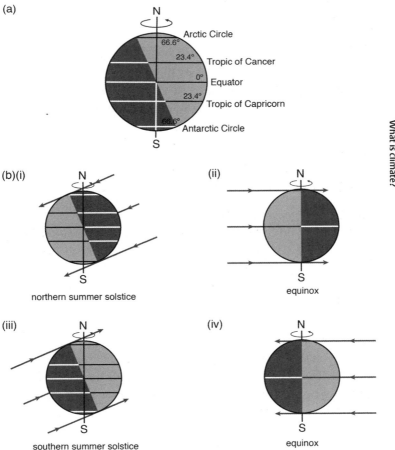

(a)

N

Arctic Circle
66.6°
23.4° Tropic of Cancer
0° Equator
23.4° Tropic of Capricorn
66.6° Antarctic Circle

S

(b)(i)
N
S
northern summer solstice

(ii)
N
S
equinox

(iii)
N
S
southern summer solstice

(iv)
N
S
equinox

4. Solstice and equinox caused by the tilt of the Earth

If we follow the sun through a year we can see how this tilt affects the Earth through the seasons. If we start at 21 June the sun is overhead at midday at the Tropic of Cancer (23.4°N), the northern summer solstice. The angle of the sun moves southward until 21 September when it is overhead at midday over the Equator, the equinox or autumn equinox in the Northern Hemisphere. The sun appears to continue southward and on 21 December it is overhead at midday at the Tropic of Capricorn .(23.4°S) the southern summer solstice. The sun then appears to move northward until it is directly overhead at midday at the Equator on the 21 March the equinox or spring equinox in the Northern Hemisphere and so the cycle continues (see Figure 4).

The seasons signal by far the most dramatic change in our climate; if we take for example New York, winter temperatures can be as low as −20°C while summer temperatures can be over 35°C—a 55°C temperature difference. Moreover as we will find out the seasons are one of the major reasons for storms.

Moving heat around the Earth

The second big factor affecting the climate of the Earth is its daily rotation. First this plunges the Earth in and out of darkness causing massive changes in diurnal temperature. For example the Sahara desert during summer can have daytime temperatures of over 38°C (100°F) and then nighttime lows of 5°C (40°F); while Hong Kong has a diurnal temperature range of little more than 4°C (7°F). Depending on the season, different areas also get varying amounts of daylight. The days can vary from 24 hours' daylight to 24 hours' darkness at the poles to around 12 hours' sunlight every day at the Equator. This change in the daylight compounds the seasonal contrasts, because not only during summer do you get more direct 'overhead' sunlight but also you get it for much longer.

But the spinning of the Earth also makes the transport of heat away from the Equator more complicated. This is because the spinning of the Earth makes everything else including the atmosphere and oceans turn. The simple rule is that rotation of the Earth causes the air and ocean currents to be pushed to the right of the direction they are travelling in the Northern Hemisphere and to the left of the direction they are travelling in the Southern Hemisphere. This deflection is called the Coriolis effect and its strength increases the further you go towards the poles.

An everyday example of this, which is always quoted, is the way water flows down a plughole or a toilet. In the Northern Hemisphere water is said to flow clockwise down the plughole while in the Southern Hemisphere it is anti-clockwise. However, I hate to tell you that the direction the water drains out of your bath or toilet is not related to the Coriolis effect or to the rotation of the Earth. Moreover no consistent difference in rotation direction between toilets in the Northern and Southern Hemispheres has been observed. This is because the Coriolis effect has such a small influence compared with any residual movement of the water and the effect of the shape of the container. This also mean the wonderful cottage industry of communities living on the Equator showing tourists the Coriolis effect is simply done by a sleight of hand. For example in Kenya there are big signs up telling you when you are crossing the Equator; if you care to stop at the road side locals will happily pour water from a bucket into a large funnel and seeming to demonstrate clearly that it goes a different way round when you are standing on one side of the sign than when you are standing on the other. However, this change is all in the wrist and how the water is poured in, affecting which way it goes round. Still, even though it is completely fake, I love these demonstrations as it means loads of locals and tourists get to hear about the Coriolis effect!

Back to climate, so why do the ocean currents and winds have this deflection? Imagine firing a missile from the Equator directly north. Because the missile was fired from the Earth which is spinning eastward, the missile is also moving east. As the Earth

9

spins the Equator has to move fast through space to keep up with the rest, as it is the widest part of the Earth. As you go further north or south away from the Equator the surface of the Earth curves in, so it does not have to move as fast to keep up with the Equator. So in one day the Equator must move round 40,074 km (the diameter of the Earth) a speed of 1,670 km/hour, while the Tropic of Cancer (23.4°N) has to move 36,750 km, with a speed of 1,530 km/hour, and the Arctic circle (66.6°N) has to move 17,662 km so has a speed of 736 km/hour. At the North Pole there is no relative movement at all so the speed there is 0 km/hour. A practical demonstration of this is if you hold hands with a friend and stand in the same place while spinning them around, they will travel much faster than you do. Therefore the missile, fired from the Equator, has the faster eastward speed of the Equator; as it moves northward towards the Tropic of Cancer, the surface of the Earth is not moving as fast eastward as the missile. This gives the appearance that the missile is moving northeast as it is moving faster eastward than the area it is moving into. Of course the closer you get to the poles the greater this difference in speeds so the greater the deflection to the east (see Figure 5).

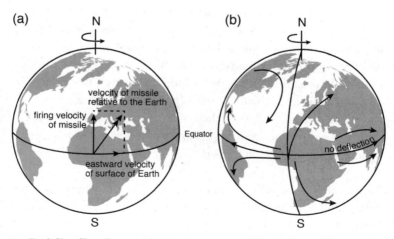

5. Coriolis effect due to relative movement of the surface of the Earth

Summary

The climate system is very straightforward. It is controlled by the different amount of solar energy received at the Equator and the poles. Climate is simply the redistribution of energy to undo this imbalance. It is the atmosphere and the oceans which undertake this redistribution, as we will see in Chapter 2. Complications are added because the Earth's axis of rotation is at an angle with respect to the sun, which leads to there being a strong season cycle. On top of this the Earth rotates every 24 hours, plunging the Earth in and out of darkness. It also means the redistribution of energy away from the Equator takes place on a spinning ball. This creates the Coriolis effect and helps to explain why nearly all weather systems seem to spin.

Chapter 2
Atmosphere and oceans

This chapter examines the effects of both the atmosphere and the oceans on climate and how they store and redistribute solar heat around the globe. It will explain why the ocean dominates in the movement of heat away from the Equator while the atmosphere dominates in the mid- to high latitudes. The chapter will finish by summarizing the major climate zones of the world and explaining why there are globally three main rain belts and two main desert belts.

The atmosphere

The atmosphere is the home of our weather. It begins at the surface of the Earth and becomes thinner and thinner with increasing altitude, with no definite boundary between the atmosphere and outer space. The arbitrary Kármán line at 62 miles (100 km), named after Theodore von Kármán (1881–1963), a Hungarian-American engineer and physicist, is usually used to mark the boundary between atmosphere and outer space. The layer of atmosphere in which weather takes place is thinner at about 10 miles thick. The oceans also play an important part in controlling our weather and climate. The oceans are on average about 2.5 miles deep, so the total thickness of the layer controlling our climate is 12.5 miles thick.

The atmosphere is a mechanical mixture of gases, not a chemical compound. What is significant is that these gases are mixed in remarkably constant proportions up to about 50 miles (80 km) above the surface of the Earth. Four gases, nitrogen, oxygen, argon, and carbon dioxide account for 99.98 per cent of air by volume. Of special interest are the greenhouse gases that despite their relative scarcity have a great effect on the thermal properties of the atmosphere, which include carbon dioxide, methane, and water vapour. The greenhouse effect is discussed later in this chapter while global warming is discussed at greater length in Chapter 8.

Content of the atmosphere

Nitrogen is a colourless, odourless, tasteless, and mostly inert gas and makes up ~78 per cent by volume of the Earth's atmosphere. Argon is also a colourless, odourless, tasteless, and completely inert gas and makes up ~0.9 per cent by volume of the Earth's atmosphere. In contrast oxygen is a very reactive gas and makes up ~21 per cent of the Earth's atmosphere by volume. Oxygen sustains all life on Earth and is constantly recycled between the atmosphere and the biological processes of plants and animals. It combines with hydrogen to produce water, which in its gaseous state, water vapour, is one of the most important components of the atmosphere as far as weather is concerned.

Oxygen also forms another gas called ozone or trioxygen, which is made up of three oxygen atoms instead of the usual two. This is an extremely important gas in the atmosphere as it forms a thin layer in the stratosphere (between 6 and 31 miles) that filters out harmful ultraviolet radiation that can cause cancer. However, even in this 'layer' the ozone concentrations are only two to eight parts per million in volume, so most of the oxygen remains of the normal dioxygen type. Much of this important gas was being destroyed by our use of CFCs, and ozone holes have been found over the Arctic and Antarctic, until governments worldwide agreed (for example, in the Vienna Convention for the Protection of the Ozone Layer in 1985 and then in the Montreal Protocol on Substances that Deplete the Ozone Layer in 1987) to stop the use of all CFCs and related compounds.

Carbon dioxide makes up 0.04 per cent of the Earth's atmosphere and is a major greenhouse gas, important for keeping the Earth relatively warm. Until recently, the level of carbon dioxide has been balanced through its consumption by plants for photosynthesis and its production by plants and animals in respiration. However, human industry over the last 100 years has caused a lot more carbon dioxide to be pumped out into the atmosphere, upsetting this natural balance.

Aerosols are suspended particles of sea salt, dust (particularly from desert regions), organic matter, and smoke. The height at which these aerosols are introduced will determine whether they cause regional warming or regional cooling. This is because high up in the atmosphere they help reflect sunlight thus cooling the local area, while at low altitudes they absorb some of the warmth coming off the Earth thus warming the local air. Industrial processes have increased the level of aerosols in the atmosphere, which has lead to smog in urban areas, acid rain, and localized cooling causing 'global dimming'. But the most important effect of aerosols is to help clouds form. Without these minute particles water vapour cannot condense and form clouds; and without cloud precipitation there is no weather.

Water vapour is the forgotten but most important greenhouse gas, which makes up about 1 per cent by volume of the atmosphere, but is highly variable in time and space as it is tied to the complex global hydrological cycle. The most important role that water vapour plays in the atmosphere is the formation of clouds and the production of precipitation (rain or snow). Warm air can hold more water vapour than cold air. So whenever a parcel of air is cooled down, for example as air rises or meets a cold air mass, it cannot hold as much water vapour, so the water condenses on to aerosols and produces clouds. An important point which we discuss later is that as water changes from a gas to a liquid it releases some energy, and it is this energy which can fuel storms as large as hurricanes. Clouds come in all sorts of shapes and sizes and are an excellent way of telling what sort of weather is coming up!

Greenhouse effect

The temperature of the Earth is determined by the balance between energy from the sun and its loss back into space. Of Earth's incoming solar short-wave radiation (mainly ultraviolet radiation and visible 'light'), nearly all of it passes through the atmosphere without interference (Figure 6). The only exception is ozone, which luckily for us absorbs energy in the high-energy UV band (which is very damaging to our cells), restricting how much reaches the surface of the Earth. About one-third of the solar energy is reflected straight back into space. The remaining energy is absorbed by both the land and ocean, which warms them up. They then radiate this acquired warmth as long-wave infrared or 'heat' radiation. Atmospheric gases such as water vapour, carbon dioxide, methane, and nitrous oxide are known as greenhouse gases as they can absorb some of this long-wave radiation, thus warming the atmosphere. This effect has been measured in the atmosphere and can be reproduced time and time again in the laboratory. We need this greenhouse effect because without it, the Earth would be at least 35 °C colder, making the average temperature in the tropics about −5 °C. Since the Industrial Revolution we have been burning fossil fuels (oil, coal, natural gas) deposited hundreds of millions years ago, releasing the carbon back into the atmosphere as carbon dioxide and methane, increasing the 'greenhouse effect', and elevating the temperature of the Earth. In effect we are releasing ancient stored sunlight back in to the climate system thus warming up the planet.

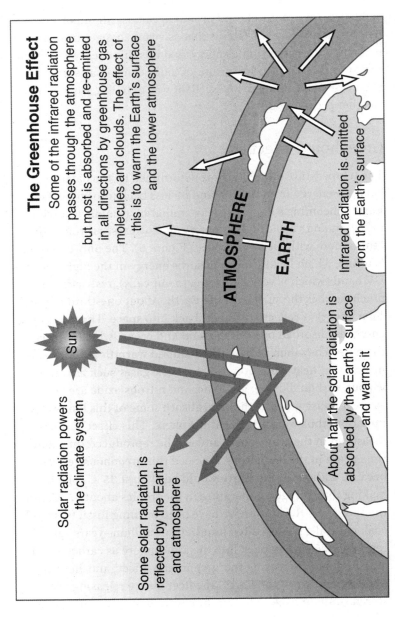

The Greenhouse Effect

Some of the infrared radiation passes through the atmosphere but most is absorbed and re-emitted in all directions by greenhouse gas molecules and clouds. The effect of this is to warm the Earth's surface and the lower atmosphere

Solar radiation powers the climate system

Some solar radiation is reflected by the Earth and atmosphere

ATMOSPHERE

EARTH

About half the solar radiation is absorbed by the Earth's surface and warms it

Infrared radiation is emitted from the Earth's surface

Sun

6. The greenhouse effect

Box 1 Vertical structure of the atmosphere

The atmosphere can be divided conveniently into a number of well demarcated horizons, mainly based on temperature (see Figure 7).

Troposphere

The lowest layer of the atmosphere is the zone where atmospheric turbulence and weather are most marked. It contains 75 per cent of the total molecular mass of the atmosphere and virtually all the water vapour. Throughout this layer there is a general decrease in temperature at a mean rate of 6.5°C/km, and the whole zone is

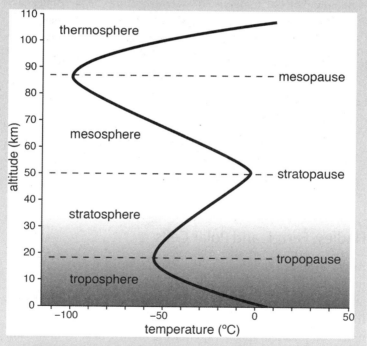

7. **Temperature profile through the atmosphere**

capped by a temperature inversion layer. This layer, called the 'tropopause', acts as a lid on the troposphere and on weather.

Stratosphere

The second major atmospheric layer extends upwards from the tropopause to about 50 km. Although the stratosphere contains much of the ozone, the maximum temperature caused by the absorption of ultraviolet radiation occurs at the 'stratopause' where temperatures may exceed 0°C. This large temperature increase is due to the relative low density of the air at this height.

Mesosphere

Above the stratopause average temperatures decrease to a minimum of −90°C. Above 80 km temperatures begin rising again because of absorption of radiation by both ozone and oxygen molecules. This temperature inversion is called the 'mesopause'. Pressure is extremely low in the mesosphere decreasing from 1 mb at 50 km to 0.01 mb at 90 km (surface pressure is about 1,000 mb).

Thermosphere

Above the mesopause, atmospheric densities are very low. Temperatures rise throughout this zone due to the absorption of solar radiation by molecular and atomic oxygen.

Hadley, Ferrel, and Polar Cells

As we have seen the shape of the Earth sets up a temperature imbalance between the Equator and the poles. Both the atmosphere and the oceans act as transporters for this heat away from the Equator. But as always with climate things get a little more complicated. At the Equator the intense heat from the sun warms up the air near the surface and causes it to rise high into the atmosphere. Warm air rises because the gas molecules in warm air can move further apart making the air less dense, and correspondingly cold air sinks. This loss of air upwards creates a space and low atmospheric pressure, which is filled by air being sucked in. This produces the Trade Winds in both the North and South Hemisphers. The northeast and southeast Trade Winds meet at the Inter-Tropical Convergence Zone (ITCZ). This causes a problem as the climate system is desperately trying to export heat away from the region around the Equator and these in-blowing winds do nothing to help this removal of heat. So in the tropics it is the surface currents of the ocean that transport most of the heat (see Figure 8). These currents include the Gulf Stream, which takes heat from the tropical Atlantic and transports it northward keeping Europe's weather mild all year round (see Box 2). Other currents include the Kuroshio current in the western North Pacific, the Brazilian current in the western South Atlantic, and finally the East Australian current in the western South Pacific.

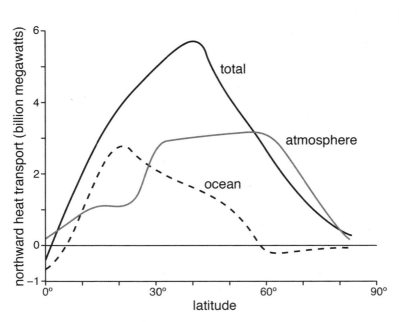

8. **Heat transport away from the Equator**

However, the hot air which has risen high into the atmosphere in the tropics slowly cools, due to both its rise and its movement towards the poles, and at about 30° north and south it sinks, forming the sub-tropical high pressure zone. As this sinking air reaches the surface it spreads out, moving both north and south. This sinking air has lost most of its moisture and therefore dries out the land it sinks onto, producing some of the major deserts around the world. The southward air links into the first atmospheric cell called the Hadley Cell and becomes part of the Trade Wind system. While the northward-bound air forms the Westerlies and it is from here northwards that the atmosphere takes over from the oceans as the major transporter of heat. The movement of warm sub-tropical air northward is only stopped when it meets the cold Polar air mass at the Polar Front. The intense cold at the poles causes air to become super chilled and sink, causing out-blowing winds (Figure 9). When this cold Polar air meets the warm, moist Westerlies at the Polar Front the clash causes the Westerlies to lose a lot of their moisture in the form of rain. It also forces the warm sub-tropical air to rise, as the cold Polar air is much heavier. This rising air completes the other two cells, the Ferrel or mid-latitude cell, and the Polar Cell—because as the air rises it spreads out to both the north and south. To the south this high-rise air meets with tropical air coming northward and sinks forming the middle Ferrel Cell. The northward component of this rising air drifts over the poles where it is chilled and sinks forming those Polar out-blowing winds which complete the third Polar Cell (Figure 9). The names of two of the three cells come from George Hadley, an English lawyer and amateur meteorologist, who in the early 18th century explained the mechanism which sustained the Trade Winds. In the mid-19th century William Ferrel, an American meteorologist, developed Hadley's theories by explaining the mid-latitude atmospheric circulation cell in detail.

An important component of these cells is the high altitude, fast flowing, narrow air currents called jet streams. The main jet streams are located near the tropopause, which represents the

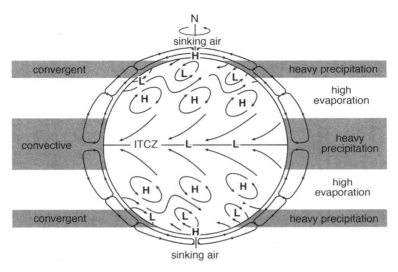

9. Major atmospheric circulation patterns

transition between the troposphere and the stratosphere (see Box 1). The major jet streams are westerly winds that flow west to east. Their paths typically have a meandering shape; jet streams may start, stop, split into two or more parts, combine into one stream, or flow in various directions including the opposite direction of most of the jet. The strongest jet streams are the polar jets, at around 7–12 km above sea level, and the higher and somewhat weaker sub-tropical jets at around 10–16 km (see Figure 10). The Northern and the Southern Hemispheres each have both a Polar jet and a sub-tropical jet. The Northern Hemisphere Polar jet flows over the middle to northern latitudes of North America, Europe, and Asia and their intervening oceans, while the Southern Hemisphere Polar jet mostly circles Antarctica all year round. Jet streams are caused by a combination of the Earth's rotation and energy in the atmosphere, hence they form near boundaries of air masses with significant differences in temperature (Figure 10).

Though the general wind patterns of Earth follow this simple three-celled, two jet stream per hemisphere model, in reality

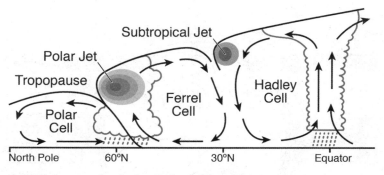

10. Major atmospheric circulation cells and jet streams

they are much more complicated. First because the Earth is spinning and this adds the influence of the Coriolis effect. This means that air masses trying to flow northward or southward are deflected by the spinning of the Earth. For example this causes large meanders in the jet streams, which are called planetary waves. These can have a huge effect on our weather, for instance in spring and summer 2012 the planetary waves within the Polar jet became fixed and brought a major heatwave to the USA and the wettest April, May, and June on record for England. Second, the continents heat up much quicker than the oceans, which can cause the surface air over the land to rise, which can alter the general circulation of surface wind. This can cause local land–sea breezes and, on a much larger scale, cause the monsoon systems. The seasons, then, can have a huge effect on atmospheric circulation, because during the summer in each hemisphere the land heats up much more than the ocean, hence the ITCZ is pulled southwards towards Australasia, and across South America and Southeast Africa during Southern Hemisphere summer and northwards across India, Southeast Asia and North Africa during Northern Hemisphere summer.

The Hadley Cells however do explain why there are three main rainfall belts across the Earth, the convection rainfall belt which moves north and south of the Equator and the two convergent rainfall belts one in the Northern and one in the Southern

Hemisphere where warm, moist sub-tropical air meets cold dry Polar air. They also explain why there are two main desert belts in the world, which are usually found between the rainfall belts with super dry air sinking between the Hadley and Ferrel Cells. In the Northern Hemisphere good examples are the Sahara desert in North Africa and the Gobi desert in China, while in the Southern Hemisphere, Central Australia and the Kalahari desert in South Africa are good examples.

The Hadley Cells can also be used to define the three main storm zones. First are 'winter storms' at the Polar Front. Second are the sub-tropical highs and the Trade Wind belt, which are the spawning ground for hurricanes. Third is the ITCZ, where the rapidly rising air cools and produces tropical thunderstorms with heavy rainfall, producing monsoons as it moves over the land (see Chapter 4 for more details).

Surface ocean circulation

As we have seen the surface ocean is important in transporting heat around the globe. The circulation of the oceans starts with the wind, because it is the action of the wind on the surface ocean that makes it move (Figure 11). As the wind blows on the surface water, the friction allows energy to be transferred from the winds to the surface water, leading to major currents. The wind energy is transferred to greater depths in the water column turbulence, which allows wind driven currents to be very deep. There are three main types of current flow: (a) Ekman motion or transport; (b) inertia currents; and (c) geostrophic currents.

Ekman motion or transport

Vagn Walfrid Ekman (3 May 1874–9 March 1954) was a Swedish oceanographer who calculated that with a constant wind over an ocean that was infinitely deep and infinitely wide with the same density, the Coriolis effect would be the only other force acting on the water column. The further away from the surface and the

Climate

11. Major surface ocean current

24

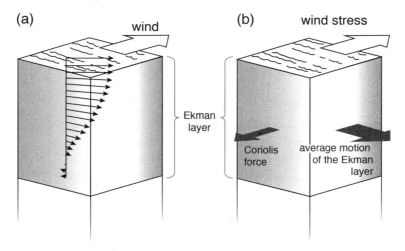

12. Ekman surface ocean movement due to wind action

diminishing influence of the wind, the greater the effect of Coriolis, which results in a spiral of water movement (see Figure 12). The result is that the net movement of the surface of the ocean is at 90 degrees to wind direction. This phenomenon was first noted by Fridtjof Nansen, during his arctic expeditions in the 1890s, when he recorded that ice transport appeared to occur at an angle to the wind direction. The direction of transport is of course dependent on the hemisphere. In the Northern Hemisphere this transport is at a 90° angle to the right of the direction of the wind, and in the Southern Hemisphere it occurs at a 90° angle to the left of the direction of the wind.

Inertia currents

Surface water masses are huge. For example, the Gulf Stream measures about 100 Sverdrup (Sv). One Sverdrup is $10^6 m^3/s$ or a million tonnes of water per second. The entire global input of freshwater from rivers to the ocean is equal to about 1 Sv. Hence these water masses have a huge momentum, and thus the currents continue to flow long after the wind has ceased pushing. When the wind stops blowing only friction and the Coriolis effect continues to act on the water mass. If the water mass does not change latitude then the current will flow along the line of latitude. If it changes latitude then the Coriolis effect acts and thus the path of the current will become even more steeply curved.

Geostrophic currents

Contrary to Ekman's assumptions, oceans are not infinitely wide and infinitely deep. The oceans have boundaries—the continents—and the water driven by the wind tends to 'pile up' on one side of the ocean against the continent. This causes a sea-surface slope, and affects the hydrostatic pressure with water flowing from areas of high to those of low pressure. This force is known as the horizontal pressure gradient force, and is also influenced by the Coriolis effect, producing what are known as geostrophic currents. One way of studying geostrophic currents is to look at the dynamic topography of the sea-surface—in other words, areas of the sea that are higher than the rest.

The combination of wind-blown Ekman currents, inertia currents, and geostrophic currents produces most of the major circulation features of the world's oceans (Figure 11). One of the major features is the gyres in each of the ocean basins. These large systems of rotating ocean currents are found in the North and South Atlantic Oceans, North and South Pacific Oceans, and the Indian Ocean. There is, however, another influence on surface ocean circulation and that is the pulling created by the sinking of surface water when deep-water currents are formed.

Deep-ocean circulation

The circulation of the deep ocean is one of the major controls on global climate due to its ability to exchange heat between the two hemispheres. In fact, the deep ocean is the only candidate for driving and sustaining internal long-term climate change (of hundreds to thousands of years) because of its volume, heat capacity, and inertia. Today the tropical sun heats the surface water in the Gulf of Mexico. This heat also causes there to be a lot of evaporation sending moisture into the atmosphere starting the hydrological cycle. All this evaporation leaves the surface water enriched in salt. So this hot salty surface water is pushed by the winds out of the Caribbean along the coast of Florida and into the North Atlantic Ocean. This is the start of the famous Gulf Stream (Figure 13). The Gulf Stream is about 500 times the size of the Amazon River at its widest point and flows along the coast of the USA and then across the North Atlantic Ocean, past the coast of Ireland, past Iceland, and up into the Nordic Seas. As the Gulf Stream flows northward it becomes the North Atlantic Drift and it cools down. The combination of a high salt content and low temperature increases the surface water density or heaviness.

Let us now examine the difference between freshwater and seawater. As freshwater is cooled down, something amazing

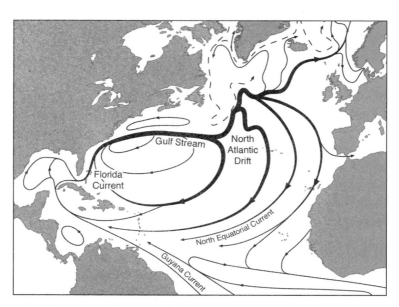

13. **Major North Atlantic surface ocean currents**

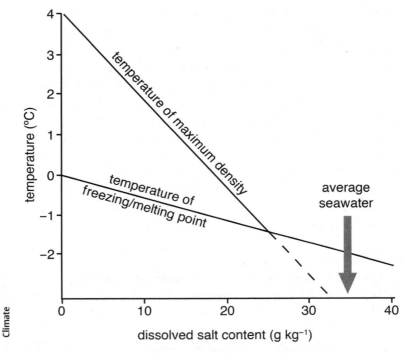

14. Temperature, salt, and density relationship for water

happens—it becomes denser down to a temperature of 4 °C, after which it becomes lighter, and then freezes at 0 °C. This means that when ponds freeze they do so from the top as the heaviest water sits on the bottom and is at 4 °C, perfect for protecting any life within the pond or lake. As you progressively add salt to water, its freezing point drops, which is why we put salt on roads to stop them freezing, but also the temperature of greatest density drops (Figure 14). At 26 grams of salt per kilogram of water the temperature of greatest density and the freezing point coincide. This means seawater, which has 35 grams of salt per kilogram, will continue to get heavier and heavier until it freezes. When water freezes then another amazing thing happens—ice is formed, a solid that is lighter than its liquid form.

When the surface water reaches the relatively fresh oceans north of Iceland, the surface water has cooled sufficiently to become dense enough to sink into the deep ocean. The 'pull' exerted by the sinking of this dense water mass helps maintain the strength of the warm Gulf Stream, ensuring a current of warm tropical water flowing into the northeast Atlantic, sending mild air masses across to the European continent. It has been calculated that the Gulf Stream delivers the same amount of energy as a million nuclear power stations. If you are in any doubt about how good the Gulf Stream is for the European climate, compare the winters at the same latitude on either side of the Atlantic Ocean, for example London with Labrador, or Lisbon with New York. Or, better still, compare Western Europe and the west coast of North America, which have a similar geographical relationship between the ocean and continent—for example, Alaska and Scotland, which are at about the same latitude.

The newly formed deep water in the Nordic Seas sinks to a depth of between 2,000 metres and 3,500 metres in the ocean and flows southward down the Atlantic Ocean, as the North Atlantic Deep Water (NADW). In the South Atlantic Ocean, it meets a second type of deep water, which is formed in the Southern Ocean and is called the Antarctic Bottom Water (AABW). This is formed in a different way to NADW. Antarctica is surrounded by sea ice and deep water forms in coast polynyas (large holes in the sea ice). Out-blowing Antarctic winds push sea ice away from the continental edge to produce these holes. The winds are so cold that they super-cool the exposed surface waters. This leads to more sea-ice formation and salt rejection because when ice is formed it rejects any salt that the freezing water contains, which produces the coldest and saltiest water in the world. AABW flows around the Antarctic and penetrates the North Atlantic, flowing under the warmer and thus somewhat lighter NADW. The AABW also flows into both the Indian and Pacific Oceans. The NADW and AABW make up the key elements of the great global ocean conveyor belt (Figure 15), which allows heat to be exchanged

29

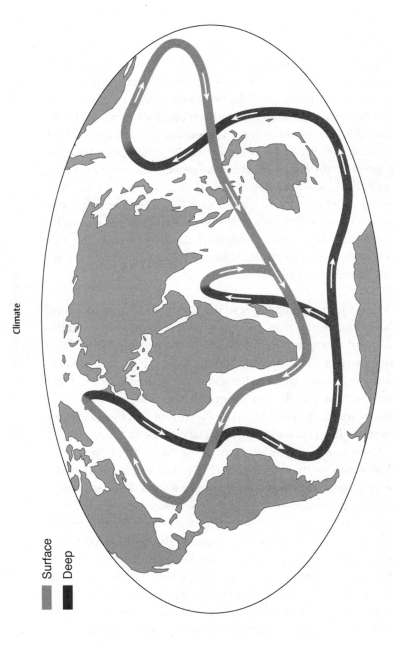

15. Global deep-ocean conveyor belt circulation

Surface
Deep

between the two hemispheres on the timescale of hundreds and thousands of years.

The balance between the NADW and AABW is extremely important in maintaining our present climate, as not only does it

Box 2 Blond hair and ocean circulation

The Gulf Stream may have also given us blond, fair skinned people. The warming effect of the Gulf Stream on Western Europe is so great that it means that early agriculturalists could grow crops incredibly far north in countries such as Norway and Sweden. These early settlers were living as far north as the Arctic Circle, which is on the same latitude as the middle of the Greenland ice sheet or the northern Alaska tundra. But there is one major drawback to living so far north and that is the lack of sunlight. Humans need Vitamin D, without it children develop rickets, which causes softening of the bones, leading to fractures and severe deformity. Vitamin D is made in the skin when it is exposed to ultraviolet light from the sun. This of course was no problem for our ancestors who evolved in Africa—quite the reverse, and dark skin was essential protection from the strong sunlight. However, as our ancestors moved further and further north there was less and less sunlight and less production of Vitamin D. In each generation only those with the lightest skin and hair colour could avoid getting rickets since the lighter your skin and hair, the more sunlight you can absorb, and thus the more Vitamin D you can make. So there was a very strong selection pressure in these areas in favour of fair skinned, blond haired people. On the other hand, Vitamin D is also found in food, such as fatty fish species and mushrooms, which may be why the same selection pressure did not apply to the Arctic Inuit. However, it is interesting to think that if it were not for the Gulf Stream and the stubbornness of the early Scandinavian settlers, relying only on crops and eating little or no fish, we would not have real blonds.

keep the Gulf Stream flowing past Europe, but it maintains the right amount of heat exchange between the Northern and Southern Hemispheres. Scientists are worried that the circulation of deep water could be weakened or 'switched off' if there is sufficient input of fresh water to make the surface water too light to sink. This evidence has come from both computer models and the study of past climates. Scientists have coined the phrase 'dedensification' to mean the removal of density by adding fresh water and/or warming up the water, both of which prevent seawater from being dense enough to sink. There is concern that climate change could cause parts of Greenland to melt. This could lead to more fresh water being added to the Nordic seas, thereby weakening the NADW and the Gulf Stream. This would bring much colder European winters with generally more severe weather. However, since the influence of the warm Gulf Stream is mainly in the winter, this change would not affect summer temperatures. So, if the Gulf Stream fails, global warming would still cause European summers to heat up. Europe would end up with extreme seasonal weather very similar to that of Alaska.

Global vegetation

The vegetation zones of the world are controlled by the annual average and seasonality of both temperature and precipitation. Temperature follows a latitudinal gradient with warmest conditions in the tropics and coldest at the poles. As we have seen there are three main rainfall belts, the convection rainfall belt in the tropics and the convergent rainfall belt in the mid-latitudes of the Northern and Southern Hemispheres. The two main desert regions of the world lie between these rainfall belts. Vegetation follows these climate zones. So rainforest is found in the tropics where there is a lot of rainfall all year round. Savannah is found in the tropics when rainfall is high seasonally, but there are also long dry seasons lasting over 4 months. The world's largest deserts are found in the mid-latitudes. Here the seasonality of rainfall is critical, as while many deserts have the same rainfall as, say,

London, this rain falls over a very small period of time, with the rest of the year being extremely arid. When the rainfall occurs only in the winter months followed by a very dry summer period, the unique Mediterranean flora is found, such as in California, South Africa, and of course around the Mediterranean. In high mid-latitudes are the temperate or boreal forests. In areas with low annual rainfall, steppe vegetation is found. In high latitudes where the temperature is the limiting factor, tundra is found. Other factors can influence where different vegetation can exist, for example we have seen that major ocean currents can allow temperate-weather vegetation to exist much further north than would usually be expected. In Chapter 5 we will see that mountain ranges and plateaus have a huge influence on where rainfall occurs and thus where deserts form.

Finally, it should be remembered that vegetation has its own influence on climate. First, vegetation changes the albedo of any area, so tropical rainforests absorb much more solar radiation than does tundra. Second, vegetation is very good at recycling water and maintaining a moist atmosphere. For example, 50 per cent of all the rainfall in the Amazon Basin comes from water recycled by the trees, evaporating and creating new clouds.

Chapter 3
Weather versus climate

Introduction

Many people get weather and climate confused. This confusion is exacerbated when scientists are asked to predict climate 50 years from now when everyone knows they cannot predict the weather a few weeks ahead. So climate is generally defined as 'the average weather'. The original definition of climate was 'the average weather over 30 years', this has been changed because we now know that our climate is changing and significant changes have been seen every decade for the last 50 years. The chaotic nature of the weather can make it unpredictable beyond a few days, while understanding the climate and modelling climate change is much easier as you are dealing with long-term averages. A good comparison is that though it is impossible to predict at what age any particular person will die, we can say with a high degree of confidence that the average life expectancy of a person in a developed country is about 80. The other confusion is that people always remember extreme weather events and not the average weather. So for example everyone remembers the heatwaves in the UK in 2003 and the USA in 2012, or the floods in Pakistan and Australia in 2010. So our perception of weather is skewed by these events rather than by an appreciation of the average weather or climate.

Chaos theory

The National Weather Service in the USA spends over $1 billion per year ensuring the country has the most accurate weather prediction possible. In other countries similar resources are poured into weather agencies, as predicting the weather is big business and getting a storm prediction right can save billions of dollars and many lives. Today three–four day forecasts are as accurate as the two-day forecasts were 20 years ago. Predictions of rain in three days' time are as accurate as one-day forecasts were in the mid-1980s. The accuracy of flash flood forecasts has improved from 60 per cent correct to 86 per cent, moreover potential victims of these floods get nearly an hour's warning instead of the 8 minutes they would have had in 1986. The lead times of advance warnings of tornadoes, in other words, the time that residents have to react, has increased from 5 minutes in 1986 to over 12 minutes. Severe local thunderstorms and similar cloudbursts are typically seen 18 minutes beforehand rather than 12 minutes over two decades ago. Seventy per cent of all hurricane paths can be predicted at least 24 hours in advance and the landfall of a hurricane can be predicted to within 100 miles (160 km).

These are great achievements but it does not explain why with all our technology and our understanding of the climate system we cannot predict the weather 10 days, a month, or a year in advance. Moreover, think of all those times that the weather report on television has said it will be sunny today and then it rains. So why is it so difficult to predict the weather? In the 1950s and 1960s it was thought that our weather prediction was limited by our lack of data and that if we could measure things more accurately and clearly understand the fundamental processes we would be able to achieve a much higher level of prediction. But in 1961 Edward Lorenz a meteorologist at Massachusetts Institute of Technology made a cup of coffee that radically changed the way we think about natural systems. In 1960 Lorenz had produced one of the first computer models of weather. One day in the winter of 1961, Lorenz's computer model produced some very interesting patterns, which he wanted to look at in greater detail. So he took a short-cut and started mid-way through the run. Of course this was one of the earliest computers so he had to retype all the starting numbers. Instead of typing them into six decimal places (e.g., 0.506127) he only typed the first three to save time and space, and then went and made the famous coffee. When Lorenz came back he found that the weather patterns had diverged from the initial run so much that there was no recognizable similarity between them. It seems the model was very sensitive to the very small changes, that one part in a thousand instead of being inconsequential had had a huge effect on the outcome. This original work has lead to the development of chaos theory. Chaos theory shows us that very small variations in atmosphere temperature, pressure, and humidity can have a major and unpredictable or chaotic effect on large-scale weather patterns.

Nevertheless, chaos theory does not mean there is a complete lack of order within a system. Far from it, chaos theory tells us that we can predict within certain boundaries what the weather will be like: we all know, for example, that most tornadoes occur in May in the USA and that winters are wet in England. But when it comes to more detailed prediction everything breaks down due to what has become known as the 'butterfly effect'. The idea is that small changes represented by the flapping of the wings of a butterfly can have a large effect on the weather, for example altering the strength and direction of a hurricane. As errors and uncertainties multiply and cascade upward through the chain of turbulent features from dust devils and squalls up to continental size eddies that only satellites can see. In effect, we will never know which of the small weather changes will combine to have these large effects. While Lorenz used 12 equations in his weather model, modern weather computers use 500,000. But even the best forecasts, which come from the European Centre for Medium Weather Forecasts based at Reading in England, suggest that weather predictions for more than four days are at best speculative and beyond a week worthless, all because of chaos.

So chaos theory says that we can understand weather and we can predict general changes but it is very difficult to predict individual events such as rain storms and heatwaves. The study of climate, however, has one great advantage over meteorology because it only examines averages and thus chaos theory does not affect it. Moreover when it comes to modelling future climate change we can now understand that an increase in the Earth's average temperature will make some weather phenomena more frequent and intense for example heatwaves and heavy rainfall events, while others will become less frequent and intense, for example extreme cold events and snow fall.

Decadal and quasi-periodic climate systems

The climate system contains many cycles and oscillations that complicate our ability to predict the weather. These include decadal cycles such as the North Atlantic oscillation (NAO), the Atlantic multi-decadal oscillation (AMO), Arctic oscillation (AO), and the Pacific decadal oscillation (PDO). So the first of these is the NAO, which was first described in the 1920s by Sir Gilbert Walker (14 June 1868–4 November 1958), a British physicist and statistician. The NAO is a climate phenomenon in the North Atlantic Ocean and is represented by the atmospheric pressure difference at sea level between Iceland and the Azores. The difference in Icelandic low-pressure and the Azores high-pressure systems seems to control the strength and direction of westerly winds and storm tracks across the North Atlantic Ocean. This in turn controls where and when in Europe it rains. Unlike the El Niño–Southern Oscillation, the NAO is largely controlled by changes in the atmosphere. The NAO is closely related to the AO and though both seem to change on a decadal scale there seems to be no periodicity. The NAO should not, however, be confused with the AMO.

The AMO is the decadal-scale variability in the sea-surface temperatures of the North Atlantic Ocean. Over the last 130 years, 1885–1900, 1927–47, 1951–61, 1998–present day, the North Atlantic Ocean temperatures have been warmer than average and the time in between colder. The AMO does affect air temperatures and rainfall over much of the Northern Hemisphere, in particular North America and Europe, for example the North Eastern Brazilian and African Sahel rainfall and North American and European summer climates. It is also associated with changes in the frequency of North American droughts and it may influence the frequency of severe Atlantic hurricanes. There are also irregular or quasi-periodic cycles such as the Indian Ocean Dipole and El Niño–Southern Oscillation (ENSO). Of these ENSO is by far the best known and is discussed in more detail below.

El Niño–Southern Oscillation

One of the most important and mysterious elements in global climate is the periodic switching of direction and intensity of ocean currents and winds in the Pacific. Originally known as El Niño ('Christ child' in Spanish) as it usually appears at Christmas, and now more often referred to as ENSO (El Niño–Southern Oscillation), this phenomenon typically occurs every 3 to 7 years. It may last from several months to more than a year. ENSO is an oscillation between three climates: the 'normal' conditions, La Niña, and 'El Niño'. ENSO has been linked to changes in the monsoon, storm patterns, and occurrence of droughts throughout the world. For example the prolonged ENSO event, in 1997 to 1998, caused severe climate changes all over the Earth including droughts in East Africa, northern India, north-east Brazil, Australia, Indonesia, and Southern USA; and heavy rains in California, parts of South America, the Pacific, Sri Lanka, and east central Africa. The state of the ENSO has also been linked into the position and occurrence of hurricanes in the Atlantic Ocean. For example, it is thought that the poor prediction of where Hurricane Mitch (see Chapter 4) made landfall was because the ENSO

38

conditions were not considered and the strong Trade Winds helped drag the storm south across central USA instead of west as predicted.

An El Niño event is when the warm surface water in the western Pacific moves eastward across to the centre of the Pacific Ocean (Figure 16). Hence the strong convection cell or warm column of rising air is much closer to South America. Consequently the Trade Winds are much weaker and the ocean currents crossing the Pacific Ocean are weakened. This reduces the amount of the cold, nutrient-rich upwelling off the coast of South America and without those nutrients the amount of life in the ocean is reduced and fish catches are dramatically reduced. This massive shift in ocean currents and the position of the rising warm air changes the direction of the jet streams that upset the weather in North America, Africa, and the rest of the world. However if you ask what causes El Niño, then the answer is of the chicken and egg variety. Does the westward ocean current across the Pacific reduce in strength, allowing the warm pool to spread eastward and moving with it the wind system. Or does the wind system relax in strength, reducing the ocean currents, and allowing the warm pool to move eastwards? Many scientists believe that long period waves in the Pacific Ocean that move between South America and Australia over time help shift the ocean currents which produce either an El Niño or a La Niña period.

La Niña is a more extreme version of the 'normal' conditions. Under normal conditions the Pacific warm pool is in the western Pacific and there are strong westerly winds and ocean currents keeping it there. This results in upwelling off the coast of South America, providing lots of nutrients and thus creating excellent conditions for fishing. During a La Niña period the temperature difference between the western and eastern Pacific becomes extreme and the westerly winds and ocean currents are enhanced. La Niña impacts on the world's weather are less predictable than those of El Niño. This is because during an El Niño period the

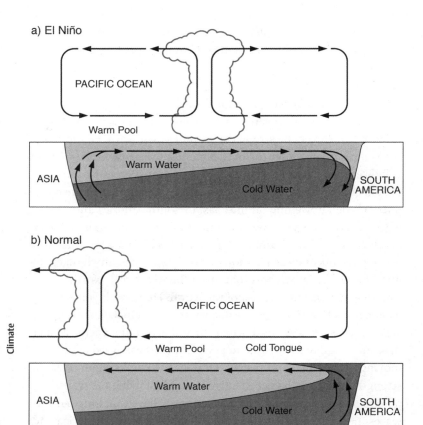

a) El Niño

PACIFIC OCEAN

Warm Pool

Warm Water

ASIA

Cold Water

SOUTH
AMERICA

b) Normal

PACIFIC OCEAN

Warm Pool   Cold Tongue

Warm Water

ASIA

Cold Water

SOUTH
AMERICA

16. Pacific Ocean during El Niño and normal periods

Pacific jet stream and storm tracks become strong and straighter
and it is therefore easier to predict its effects. La Niña on the other
hand weakens the jet stream and storm tracks, making them more
loopy and irregular, meaning that the behaviour of the atmosphere
and particularly of storms becomes more difficult to predict. In
general where El Niño is warm, La Niña is cool, where El Niño is
wet, La Niña is dry. La Niñas have occurred in 1904, 1908, 1910,
1916, 1924, 1928, 1938, 1950, 1955, 1964, 1970, 1973, 1975, 1988,
1995, 1999, 2008, and 2011, with the 2010–11 La Niña being one
of the strongest ever observed.

Predicting ENSO

Predicting an El Niño events is difficult but a lot of work has gone on for the last three decades to better understand the climate system. For example, there is now a large network of both ocean and satellite monitoring systems over the Pacific Ocean, primarily aimed at recording sea-surface temperature, which is the major indicator of the state of the ENSO. By using this climatic data in both computer circulation models and statistical models, predictions are made of the likelihood of an El Niño or La Niña event. We are really still in the infancy stage of developing our understanding and predictive capabilities of the ENSO phenomenon.

There is also considerable debate over whether ENSO has been affected by global warming. The El Niño conditions generally occur every 3 to 7 years; however, in the last 20 years, they have behaved very strangely, returning for 3 years out of 4: 1991–2, 1993–4, and 1994–5, then not returning until 1997–8, and then not returning for 8 years, finally arriving in 2006–7. Reconstruction of past climate using coral reefs in the western Pacific shows sea-surface temperature variations dating back 150 years, well beyond our historical records. The sea-surface temperature shows the shifts in ocean current, which accompany shifts in the ENSO and reveal that there have been two major changes in the frequency and intensity of El Niño events. First was a shift at the beginning of the 20th century from a 10–15-year cycle to a 3–5-year cycle. The second was a sharp threshold in 1976 when a marked shift to more intense and even more frequent El Niño events occurred. Moreover during the last few decades the number of El Niño events has increased, and the number of La Niña events has decreased. Even taking into account the effect of decadal cycles on ENSO the size of the ENSO variability in the observed data seems to have increased by 60 per cent in the last 50 years.

However, as we have seen, to predict an El Niño event 6 months from now is hard enough, without trying to assess whether or not

ENSO is going to become more extreme over the next 100 years. Most computer models of ENSO in the future are inconclusive; some have found an increase and others have found no change. This is, therefore, one part of the climate system that we do not know how global warming will affect. Not only does ENSO have a direct impact on global climate but it also affects the numbers, intensity, and pathways of hurricanes and cyclones, and the strength and timing of the Asian monsoon. Hence, when modelling the potential impacts of global warming, one of the largest unknowns is the variation of ENSO and its knock-on effects on the rest of the global climate system.

Modelling climate

The whole of human society operates on knowing the future weather. For example, a farmer in India knows when the monsoon rains will come next year and so they know when to plant the crops. While a farmer in Indonesia knows there are two monsoon rains each year so each year they can have two harvests. This is based on their knowledge of the past, as the monsoons have always come at about the same time each year in living memory. But weather prediction goes deeper than this as it influences every part of our lives. Our houses, roads, railways, airports, offices, cars, trains, and so on are all designed for our local climate. Predicting future climate is, therefore, essential as we know that global warming is changing the rules. This means that the past weather of an area cannot be relied upon to tell you what the weather in the future will hold. So we have to develop new ways of predicting and modelling the future, so that we can plan our lives and so that human society can continue to fully function.

There is a whole hierarchy of climate models, from relatively simple box models to the extremely complex three-dimensional general circulation models (GCMs). Each has a role in examining and furthering our understanding of the global climate system. However, it is the complex three-dimensional general circulation models which are used to predict future global climate. These comprehensive climate models are based on physical laws represented by mathematical equations that are solved using a three-dimensional grid over the globe. To obtain the most realistic simulations, all the major parts of the climate system must be represented in sub-models, including the atmosphere, ocean, land surface (topography), cryosphere, and biosphere, as well as the processes that go on within them and between them. Most global climate models have at least some representation of each of these components. Models that couple together both the ocean and atmosphere components are called atmosphere–ocean general circulation models (AOGCMs).

Over the last 25 years there has been a huge improvement in climate models. This has been due to our increased knowledge of the climate system but also because of the nearly exponential growth in computer power. There has been a massive improvement in spatial resolution of the models from the very first Intergovernmental Panel on Climate Change (IPCC) in 1990 to the latest in 2007. The current generation of AOGCMs has a resolution of one point every 110 km by 110 km, and this is set to get even finer when the next IPCC Science Report is published in late 2013. The very latest models or as some groups are now referring to them 'climate simulators' include much better representations of atmospheric chemistry, clouds, aerosol processes, and the carbon cycle including land vegetation feedbacks. But the biggest unknown or error in the models, is not the physics, it is the estimation of future global greenhouse emissions over the next 90 years. This includes many variables, such as the global economy, global and regional population growth, development of technology, energy use and intensity, political agreements, and personal lifestyles.

Over 20 completely independent AOGCMs have been run using selected future carbon dioxide emission scenarios for the IPCC 2007 report, producing global average temperature changes that may occur by 2100. This is a significant change from the IPCC 2001 report, in which only 7 of these models were used. Using the widest range of potential emission scenarios the climate models suggest that global mean surface temperature could rise by between 1.1°C and 6.4°C by 2100. Using the best estimates for the 6 most likely emission scenarios, then this range is 1.8°C to 4°C by 2100. Model experiments show that even if all radiation forcing agents were held at a year-2000 constant, there would still be an increase of 0.1°C per decade over the next 20 years. This is mainly due to the slow response of the ocean. Interestingly, the choice of emission scenario has little effect on the temperature rise to 2030, making this a very robust estimate. All models suggest twice the rate of temperature increase in the next two decades compared with that of the 20th century. What is significant is that the choices we make now in terms of global emissions will have a significant effect on global warming after 2030. The next IPCC report to be published in late 2013, though it will use greatly improved emission scenarios, will have a very similar potential change of warming by the end of the century. What is amazing and very reassuring is that over the last 25 years the climate models have consistently given us the same answer, meaning we do understand the climate system and we can understand the consequences of our past and future actions.

Chapter 4
Extreme climates

Introduction

Humans can live, survive, and even flourish in the extreme climates ranging from that of the Arctic to that of the Sahara. We have populated every continent except Antarctica. We can deal with the average climate of each region through our adaptations of technology and lifestyle. The problems arise when the predictable boundaries of local climate are exceeded, for example by heatwaves, storms, droughts, and/or floods. This means that what we define as extreme weather, such as a heatwave, in one region may be considered fairly normal weather in another. Each society has a coping range, a range of weather with which it can deal: what is seen as a heatwave in England would be normal summer conditions in Kenya. However, one of the most unpredictable and dangerous elements in our climate systems are storms. In this chapter we examine how and why storms are formed and their impact. Hurricanes, tornadoes, winter storms, and the monsoons will all be discussed.

Hurricanes

A hurricane is a severe cyclonic tropical storm that starts in the North Atlantic Ocean, Caribbean Sea, Gulf of Mexico, west coast of Mexico, or the northeast Pacific Ocean (Figure 17). They are called typhoons in the western Pacific and simply tropical cyclones in the Indian Ocean and Australasia. They are, however, all the exactly the same type of storm and here we will call all of them hurricanes. Hurricanes occur in the tropics between 30°N and 30°S, but not near the Equator as there is not enough atmospheric variation to generate them. For a storm to be classified as a hurricane, the sustained wind speed must exceed 120 km/hr. Of course in a fully developed hurricane, wind speeds can exceed 200 km/hr.

A hurricane is a tropical storm run amok, a rotating mass of thunderstorms that has become highly organized into circular cells, which are ventilated by bands of roaring winds. Hurricanes develop over the oceans and tend to lose their force once they move over land—this is because unlike temperate storms hurricanes are driven by the latent heat from the condensation of water. The sun is most intense close to the Equator where it heats the land, which in turn heats the air. This hot air rises and consequently sucks air from both the north and south producing the Trade Winds. As the seasons change so does the position of the clash of the Trade Winds, which is called the Inter-tropical Convergence Zone (ITCZ). To generate a hurricane the sea temperature must be above 26°C for at least 60 m below the surface and the air humidity must be at about 75–80 per cent. This combination provides the right amount of heat and water vapour to sustain the storm once it has started. For example these conditions occur during the summer in the North Hemisphere when the tropical North Atlantic Ocean heats up enough and its water starts to evaporate. Initially the warm ocean heats the air above it and causes that to rise. This produces a low-pressure area which sucks in air from the surrounding area. This rising air contains a lot of water vapour due to pronounced evaporation from the hot surface of the ocean. As the air rises it cools and can no longer hold as much water vapour; as a result some of it condenses to form water droplets and then clouds. This transformation from water vapour to water droplets releases energy called 'latent heat'. This in turn causes further warming of

Climate

46

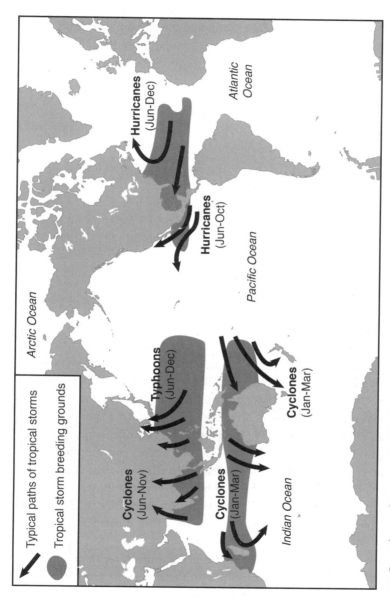

Arctic Ocean

Atlantic Ocean

Pacific Ocean

Indian Ocean

Hurricanes
(Jun–Dec)

Hurricanes
(Jun–Oct)

Typhoons
(Jun–Dec)

Cyclones
(Jun–Nov)

Cyclones
(Jan–Mar)

Cyclones
(Jan–Mar)

Typical paths of tropical storms

Tropical storm breeding grounds

17. Location and occurrence of major tropical storms

the air and causes it to rise even higher. This feedback can make the air within a hurricane rise to over 10,000 m above the ocean. This becomes the eye of the storm and the spiraling rising air it produces creates a huge column of cumulo-nimbus clouds. You can see a mini version of this with steam coming out of a kettle. As the hot air rises from the kettle it hits the colder air and it forms steam, a mini-cloud. If you have ever put your hand near the steam you can feel it is very hot and this is because of all the energy being released as the water vapour changes from a gas back to a liquid.

When the air inside the hurricane reaches its highest level it flows outwards from the eye producing a broad canopy of cirrus cloud. The air cools and falls back to sea level where it is sucked back into the centre of the storm. Because of the Coriolis force, the air that is sucked into the bottom of the hurricane spins into the storm in a clockwise direction, while the air escaping at the top spins out in a counter-clockwise direction. This pattern is the opposite in the Southern Hemisphere. Hurricanes form at least 345 miles or 5° of latitude away from the Equator, where the Coriolis effect is strong enough to give the required twist to the storm. The size of hurricanes can vary from 100 km to over 1,500 km. A hurricane can form gradually over a few days or in the space of 6 to 12 hours and typically the hurricane stage will last 2–3 days and take about 4–5 days to die out. Scientists estimate that a tropical cyclone releases heat energy at the rate of 50 to 200 exajoules (10^{18} J) per day, equivalent to about 1 PW (10^{15} Watt). This rate of energy release is equivalent to 70 times the human world energy consumption and 200 times the worldwide electrical generating capacity, or to exploding a 10-megaton nuclear bomb every 20 minutes. Hurricanes are measured in the Saffir-Simpson scale and go from a tropical storm through category 1 to the worst at category 5.

However, the formation of hurricanes is much rarer than might be expected given the opportunities for them to occur. Only 10 per cent of centres of falling pressure over the tropical oceans give rise to fully fledged hurricanes. In a year of high incidence, perhaps a

Climate

maximum of 50 tropical storms will develop to hurricane levels. Predicting the level of a disaster is difficult as the number of hurricanes does not matter—it is whether they make landfall (Figure 18). For example, 1992 was a very quiet year for hurricanes in the North Atlantic Ocean. However, in August, one of the few hurricanes that year, Hurricane Andrew, hit the USA just south of Miami and caused damage estimated at $26 billion. Hurricane Andrew also demonstrates that predicting where a storm will hit is equally important—if the hurricane had hit just 20 miles further north it would have hit the densely populated area of Miami City and the cost of the damage would have doubled.

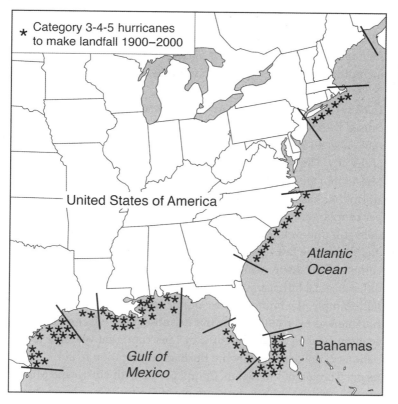

18. Major hurricane landfall over a 100-year period

In terms of where hurricanes hit in developed countries, the major effect is usually economic loss, while in developing countries the main effect is loss of life. For example, Hurricane Katrina, which hit New Orleans in 2005, caused 1,836 deaths while Hurricane Mitch, which hit Central America in 1998, killed at least 25,000 people and made 2 million others homeless. In both cases the greatest damage was caused by the huge amount of rainfall. Honduras, Nicaragua, El Salvador, and Guatemala were battered by 180-mile (290 km) per hour winds, and more than 23 inches (60 cm) of rain every day. Honduras, a small country of only 6 million inhabitants, was the worst hit. The Humuya River, normally a calm stretch of water about 200 feet (60m) wide rose by 30 feet (9 m) and became a raging torrent, ripping out trees as tall as a city block from the ground. Eighty-five per cent of the country ended up under water. Over 100 bridges, 80 per cent of the roads, and 75 per cent of its agriculture were destroyed, including most of the banana plantations.

In New Orleans the worst damage by Hurricane Katrina was caused by both the intense rainfall and the storm surge. Together they caused 53 different levees to break, submerging 80 per cent of the city. The storm surge also devastated the coasts of Mississippi and Alabama. Hurricane Katrina was not the worst storm that has hit the USA; a storm that hit Miami in 1926 was 50 per cent larger but did little damage because Miami Beach was then still undeveloped. In the USA coastal population has doubled in the last 10 to 15 years making the country much more vulnerable to storm related losses. There is also a large financial difference if a hurricane hits a developed or developing country. For example, the immediate economic impact of Hurricane Katrina was over $80 billion, but its subsequent effect on the US economy was to boost it slightly, by 1 per cent, that year due to the billions of dollars spent by the Bush administration to aid reconstruction of the region. Compare this with Hurricane Mitch in 1998, which set back the economy of Central America by about a decade.

Hurricanes also occur elsewhere in the world. An average of 31 tropical storms roam the western North Pacific every year, with typhoons smashing into Southeast Asia from June to December; most at risk are Indonesia, China, and Japan, otherwise known as 'Typhoon Alley'. Why does Typhoon Alley get so many typhoons? And why can they occur almost any time of year. The answers lie in the oceans. The key is the 'warm pool' of ocean water that sits in the western tropical Pacific. All year long the Trade Winds and the ocean current push the surface water warmed by the tropical sun to the far western side of the North Pacific. Hurricane seasons come and go in other parts of the world but the water of the 'warm pool' is always warm enough to start a hurricane—though they are most common between June and December (Figure 18).

Tornadoes

Tornadoes are nature's most violent storms. Nothing that the atmosphere can dish out is more destructive: they can sweep up anything that moves; and they can lift buildings from their foundations, making a swirling cloud of violently flying debris. They are very dangerous, not only because of the sheer power of their wind, and the missiles of debris they carry, but because of their shear unpredictability. Tornado strength and destructive capability is measured on the Fujita Scale.

A tornado is a violent rotating column of air, which at a distance appears as an ice cream cone-shaped cloud formation. Other storms similar to tornadoes in nature are whirlwinds, dust-devils (weaker cousins of tornadoes occurring in dry lands), and waterspouts (a tornado occurring over water). Tornadoes are most numerous and devastating in central, eastern, and northeastern USA, where an average of 5 per day are reported every May. They are also common in Australia (15 per year), Great Britain, Italy, Japan, Bangladesh, east India, and central Asia. While the greatest number of fatalities occurs in the United States, the deadliest tornadoes by far have occurred in the small area of Bangladesh and east India. In this 8,000 mile² (21,000 km²) area, 24 of the 42 tornadoes known to have killed more than a 100 people have occurred. This is likely due to the high population density and poor economic status of the area as well as a lack of early warning systems.

We can see tornadoes as miniature hurricanes. Although tornadoes can form over tropical oceans they are more common over land. The formation of tornadoes is encouraged when there is warm, moist air near the ground and cold dry air above. This occurs frequently in late spring and early summer over the Great Plains of the USA (Figure 19). Intense heating of the ground by the sun makes warm, moist air rise. As it does so it cools and forms large cumulo-nimbus clouds. The strength of the updraft determines how much of the surrounding air is sucked into the bottom of what becomes a tornado. Two things help the tornado to rotate violently; the first is the Coriolis force and the second is the high level jet stream passing over the top of the storm, adding an extra twist to the tornado. Because of the conditions under which tornadoes are formed they can easily occur beneath thunderstorms and hurricanes.

In the USA nearly 90 per cent of tornadoes travel from the southwest to the northeast, although some follow quick changing zigzag paths. Weak tornadoes or decaying tornadoes have a thin ropelike appearance. The most violent tornadoes have a broad dark funnel shape that extends from the dark wall cloud of a large thunderstorm. There have even been reports of some tornadoes practically standing still, hovering over a single field, and of others that crawl along at 5 miles per hour. On the other hand, some have been clocked at over 70 miles per hour. However, on average, tornadoes travel at 35 miles per hour. It has been noted that most tornadoes occur between 3pm and 9pm, but they have been known to strike at any time of day or night. They usually only last about 15 minutes, staying only a matter of seconds in any single place—but

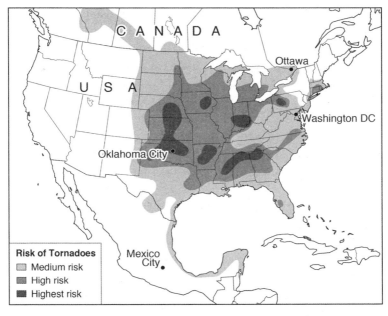

19. **USA tornado risk map**

then some tornadoes just do not fit any of these rules, for example on 18 March 1925 a single tornado travelled 219 miles in 3.5 hours through Missouri, Illinois, and Indiana killing 695 people.

Tornado Alley

Tornado Alley is the nickname for the area in which most tornadoes occur in the USA, and it expands through spring and summer as the heat from the sun grows warmer and the flow of warm moist air from the Gulf of Mexico spreads further north (see Figure 19). An area that includes central Texas, Oklahoma, and Kansas is at the hard core of Tornado Alley, but before the season is over it can have expanded to the north to Nebraska and Iowa. It shrinks and swells over time but there is only one Tornado Alley. Nowhere else in the world sees weather conditions in a combination so perfect to make tornadoes. The key reasons for this special area are: (1) beginning in spring and continuing through summer, low-level winds from the south and southeast bring a plentiful supply of warm tropical moisture up from the Gulf of Mexico into the Great Plains; (2) from down off the eastern slopes of the Rocky Mountains or from out of the deserts of northern Mexico come other flows of very dry air that travel about 3,000 feet above the ground; and (3) at 10,000 feet high the prevailing westerly winds, sometimes accompanied by a powerful jet stream, race overhead, carrying cool air from the Pacific Ocean and providing a large temperature difference, which will drive the tornadoes and the twists to get started.

In 2011 there were 1,897 tornadoes reported in Tornado Alley in the USA beating the record of 1,817 tornadoes recorded in 2004. The year 2011 was also an exceptionally destructive and deadly year in terms of tornadoes, killing at least 577 people worldwide. Of those, an estimated 553 were in the United States, which compared to 564 US deaths in the prior 10 years combined. That year saw the second greatest number of deaths due to tornadoes in a single year in US history. However, this is still a long way off from the most deadly tornado on record, which occurred on 26 April 1989 in Bangladesh and killed over 1,300 people, injured 12,000 people, and destroyed everything but a few trees from Daultipur to Salturia.

Winter storms

For people living in the mid-latitudes weather seems to be a permanent topic of conversation. This is because it is always changing. In Britain there is a saying, 'if you do not like the weather wait an hour and it will change'. This is because the climate of the mid-latitudes is dominated by the titanic clash between the cold polar air moving southward and the warm sub-tropical air moving northwards. This clash of air masses takes place at the Polar Front.

The Polar Front moves north and south with the seasons. In summer when the sub-tropical air is warmer it moves further towards the pole. During winter when conditions are much colder the polar air mass is dominant and the Polar Front moves towards the Equator. Where these two air masses meet rain is formed. This is because warm air can hold more water vapour and when it clashes with the cold air this vapour condenses into clouds, which in turn produce rain. But it is the upper atmosphere which really controls the shape and thus the weather of the Polar Front. The upper atmosphere is characterized by fast 'jet streams' that race around the planet. These powerful jet streams push the Polar Front around the Earth, but as it does so the Front wrinkles and becomes a mass of so-called planetary waves moving gradually round our planet. These waves have a great effect on our weather, causing us all to complain about the weather being so changeable and of course wet. One of these waves can pass over a town in about 24 hours. The weather will be experienced as starting out to be relatively cold but with clear skies. As the warm front passes overhead the conditions get warmer and it starts to rain—usually light rain or drizzle. As the centre of the warm air mass reaches the town the weather turns cloudy and muggy and the rain stops. Then the second front, the cold front, passes overhead; temperatures drop and there is a short period of very heavy rainfall. Then it is back to cold, clear weather until the next wave reaches the town.

As we have seen there are many storms that are associated with distinct areas of atmospheric circulation described in the section on Hadley Cells. Ice, wind, hail, and snow storms are associated with either the Polar Front or high mountain regions and are worse in the winter time. In the Northern Hemisphere these types of storms are common over North America, Europe, Asia, and Japan.

For snow to reach the ground the temperature of the air between the base of the cloud and the ground must be below 4°C, otherwise the snowflakes melt as they travel through the air. For hailstones to form the top of the storm must be very cold. High up in the atmosphere water droplets can become super cooled to less than 0°C, which collide in the atmosphere to form ice balls or hailstones. If you cut open a hailstone you can see the layers of ice that have built-up like an onion. The stones can vary from between 2 mm and 20 cm. Their size depends on how strong the

Box 3 Caught in the cold

When your body loses the battle against the cold, it is often someone else who will notice it. This is why you should always be on the look-out for the symptoms of cold weather exposure in your companions. When the cold has started to affect you badly, you are not always the best judge of the seriousness of the problem. You still think that you are okay you just need another minute's rest. These are the signs to look out for:

- You cannot stop shivering
- You are fumbling your hands
- Your speech is slow and slurred and may even be incoherent
- You stumble and lurch as you walk
- You are drowsy and exhausted and feel the need to lie down even though you are outside
- Maybe you have rested, but cannot then get up

A person acting like this needs to get into dry clothes and a warm bed. This is because the core temperature of that person has started to drop, which is extremely dangerous for the body; if it is not stopped it will result in death. They need a warm hot water bottle, heating pad, or warm towels on their body. They need warm drinks. They do NOT need an alcoholic or caffeinated drink, as these speed up the person's heart rate, causing them to lose yet more heat; they also dehydrate the body, which hinders its recovery. Also do NOT massage or rub the person, as this again takes away heat from the body core where it is most required. The person should also always be seen by a doctor.

updraft of air is, as this determines how long they stay in the atmosphere before dropping out. The worst storm conditions are called blizzards. These combine strong winds, driving snow, ice, and hail, with air temperatures as low as −12°C and visibility less than 150 metres (see Box 3).

Monsoons

The other important area for massive rainstorms is the monsoon belt. The name monsoon comes from the Arabic word 'mausim' which means 'season', as most of the rains that fall in Southeast Asia occur during the summer. In the tropics the sun's energy is most intense as the sun is directly overhead. This heats up the land and sea and thus warms the air above. This warm, moist air rises, leaving an area of low pressure beneath it, which helps to suck in air from the surrounding area (Figure 20). This suction results in the Trade Winds, which can travel from much higher latitudes to this area of rising air. As the winds come from both the Northern and Southern Hemispheres this area is known as the ITCZ. As the air at the ITCZ rises, it forms huge towering clouds and produces large amounts of rain. The ITCZ moves north and south with the seasons as the position of the most intense sunlight shifts up and down across the Equator. It is also influenced strongly by the position of the continents. This is because the land heats up faster and to a greater extent than the ocean and thus it can pull the ITCZ even further north or south during that season. An example of this is the Asian summer monsoon, during the summer near the Himalayan mountains and the low lands of India heat up. This pulls the ITCZ across the Equator on to Asia. Because the Southern Hemisphere winds have been pulled across the warm Indian Ocean they are warm and full of moisture; when they are forced to rise and cool down over India they produce very heavy rainfall throughout Southeast Asia and as far north as Japan. During Northern Hemisphere winter the ITCZ moves south of the Equator, but in Southeast Asia it means warm, moist winds from the North Pacific are dragged southward across the

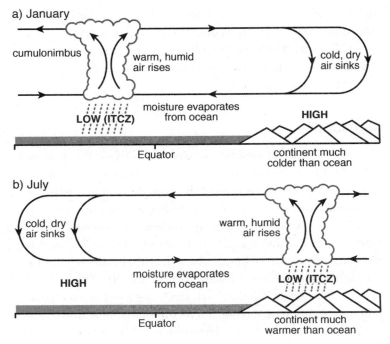

a) January

cumulonimbus

warm, humid
air rises

cold, dry
air sinks

LOW (ITCZ)

moisture evaporates
from ocean

HIGH

Equator

continent much
colder than ocean

b) July

cold, dry
air sinks

warm, humid
air rises

HIGH

moisture evaporates
from ocean

LOW (ITCZ)

Equator

continent much
warmer than ocean

Climate

20. Monsoonal system

continent into the Southern Hemisphere. This means that some areas such as Indonesia and Southern China get two monsoonal rainy seasons a year: one from the north and one from the south. No wonder this is the most fertile place on Earth, supporting over two-fifths of the world's population. Despite being the bringers of life, the rains can cause catastrophic hazards, especially in the form of flooding. Examples of this are the terrible floods in 1998 in Bangladesh and China which caused over $30 billion of damage and thousands of deaths.

Amazon monsoon

During the Southern Hemisphere summer the continent of South America heats up. This rising air leaves an area of low pressure at ground level, which is filled by sucking in the surrounding air.

This pulls the convergence zone between the North and Southern tropical air southward over Brazil. The southward shift of the ITCZ brings with it lots of rain as the air being pulled across the Equator from the north originates over the warm, tropical Atlantic Ocean. This produces the Amazon monsoon and results in the mightiest river in the world and the greatest extent of rainforest on the planet. The Amazon Basin covers an amazing 2.7 million miles^2 much of which is covered with rainforest. The Amazon River delivers 20 per cent of all the freshwater that enters the world's oceans. Without the monsoon rains the most diverse habitat in the world would not exist.

Living under the Asian monsoons

Bangladesh is a country literally built by the monsoons as over three-quarters of the country is a deltaic region formed by the sediments brought in by the Ganges, Brahmaputra, and Meghna rivers—all fed by the summer monsoons. Over half the country lies less than 5 metres above sea level, thus flooding is a common occurrence. During a normal summer monsoon a quarter of the country is flooded. Yet these floods, like those of the Nile, bring life with them as well as destruction. The water irrigates and the silt fertilizes the land. The fertile Bengal Delta supports one of the world's most dense populations, over 110 million people in 140,000 km^2. But every so often the monsoon floods exceed what even Bangladesh can cope with. In 1998 three-quarters of the country was flooded for 2 months, causing billions of pounds worth of damage and thousands of deaths. Bangladesh also has to cope with tropical cyclones. If we take 3 of the worst years for tropical cyclones we can see the loss of life has dramatically dropped. In 1970 there were over 300,000 cyclone-related deaths, in 1991 there were 138,000, while in 2007 there were just 3,500 deaths. This is not because the tropical cyclones have grown gentler, far from it. It is because of good governance. The Bangladesh government has, first, invested in excellent meteorological facilities to make as accurate a prediction of when and where the cyclones will make land fall; second, they have set up a communication network using cyclists, so that once a cyclone warning is given, the message is carried to all the towns and villages that will be affected. They have also built cyclone shelters, protected water and sanitation facilities, and encouraged floating agriculture, which can withstand the storms. These relatively simple changes have resulted in the saving of hundreds of thousands of lives.

Chapter 5
Tectonics and climate

Introduction

In Chapters 1 and 2 we saw how climate is a function of how the sun's energy falls on the Earth and is then redistributed around the globe. Both of these aspects are strongly influenced by plate tectonics. This is why 100 million years ago the Earth was much warmer and humid, and dinosaurs were happily living on Antarctica. Our modern climate system is a product of millions of years of plate tectonics, which have produced unique occurrences such as significant amounts of ice at both Poles. This produces a very strong Equator-pole temperature gradient and thus an extremely dynamic and energetic climate system. Tectonics has two main effects on climate. First, there are direct effects, which include mountain and plateau uplift which changes atmospheric circulation and the hydrological cycle or ocean gateways, which change the way the oceans circulate. Second, there are indirect effects that affect the content of the atmosphere through subduction, volcanism, and consumption of gases by chemical weathering. One of the themes running through this book is the idea that nothing in climatology is complex. This is also true of the effects of tectonics on climate. In this chapter the influences are broken down into horizontal tectonics, which examines what happens if you simply move the continental plates around the globe. Next is vertical tectonics, which examines what happens if you create a mountain or a plateau. Last, we will look at the effects of volcanoes and supervolcanoes on climate.

Horizontal tectonics

Latitudinal continents

The north–south position of the continents has a huge effect on the thermal gradient between the poles and the Equator. Geologists have run simple climate models to look at this effect (Figure 21). If you put all the continents around the Equator, the so-called 'tropical ring world', the temperature gradient between the poles and the Equator is about 30 °C (Figure 21). This is because when the poles are covered with oceans they cannot go below freezing. This is due to a trick of both the atmosphere and the oceans. A fundamental rule of climate is that hot air rises and cold air drops. At the poles it is cold so the air falls and as it hits the ground it pushes outwards away from the pole. When sea water at the pole freezes it forms sea ice, and this ice is then blown away from the pole towards warmer water where it melts. This maintains the balance and prevents the temperature of the poles going below zero. However, as soon as you introduce land onto the pole or even around the pole, ice can form permanently. If you do have a landmass like Antarctica over a pole with ice on it the Equator–pole temperature gradient is over 65 °C (Figure 21); which is exactly what we have today. In contrast if you consider the Northern Hemisphere, the continents are not on the pole but surround it. So instead of one huge ice sheet, as we have in Antarctica, there is one smaller one on Greenland, and the continents act like a fence, keeping all the sea ice in the Arctic Ocean. So the Equator–pole temperature gradient of the Northern Hemisphere is somewhere between the extremes of the Antarctic and an ice-free continent, about 50 °C. The size of the Equator–pole temperature gradient is a fundamental driver of our climate. Because the main driver of ocean and atmospheric circulation is moving heat from the Equator to the poles. So this

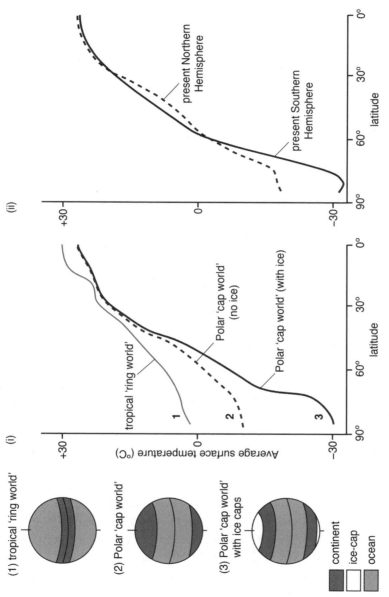

21. Latitudinal location of continent and the Equator–pole temperature gradient

temperature gradient defines what sort of climate the world will have. A cold Earth has an extreme Equator–pole temperature gradient and thus a very dynamic climate. This is why we have strong hurricanes and winter storms: the climate system is trying to pump heat away from the hot tropics towards the cold poles.

Longitude continents

Chapter 2 described the fundamentals of ocean circulation. One of the key aspects of ocean circulation is how the oceans are contained. If there are no continents in the way then oceans will just continue to circulate around and around the globe. However, when an ocean current encounters a continent it is deflected both north and south. If we look at the modern configuration of the continents (Figure 22a) then there are three main longitudinal continents: (1) the Americas, (2) Europe down to southern Africa, and (3) Northeast Asia down to Australasia. A hundred million years ago the continents are still recognizable but they are in slightly different positions (Figure 22b). The two striking features are, first, there was an ocean across the whole of the tropics through the Tethyan Sea and the Deep Central American passage. Second, there is no ocean circulating around Antarctica. These changes have huge effects on the circulation of the surface ocean and thus deep-ocean circulation and global climate. There are three main conceptual ways of understanding the effects of ocean gateways on ocean circulation. The first is a simple double-slice world with longitudinal continents on either side (Figure 23). Because ocean currents are driven by the surface winds in the tropics and poles the ocean currents are pushed to the west, while in the mid-latitudes they are pushed to the east. This produces the classic two-gyre solution in both hemispheres. Today both the North Pacific Ocean and the North Atlantic Ocean have this type of circulation. The second scenario is a double-sliced world with a low latitude seaway. This produces a large tropical ocean circulating continually westward around the world. There are then two smaller gyres in each hemisphere (Figure 23). This is the circulation seen during the

Cretaceous period, with the two gyres in each hemisphere occurring in the Pacific Ocean. The third scenario is a double sliced world with high latitude seaways. This produces strong circumpolar ocean currents in each hemisphere and a single tropical gyre in each hemisphere (Figure 23). Today the Southern Hemisphere resembles this scenario with a circumpolar current around Antarctica. The Southern Ocean thus acts like a giant ocean heat extractor and was instrumental in the huge build up of ice on Antarctica.

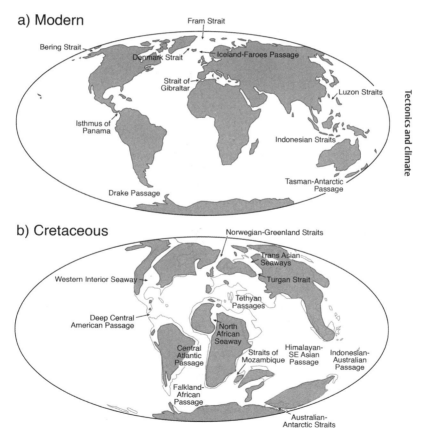

a) Modern

b) Cretaceous

22. **Ocean gateways both today and during the Cretaceous period**

N

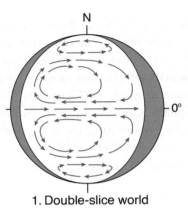

1. Double-slice world

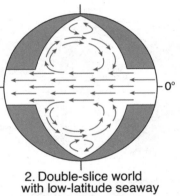

2. Double-slice world
with low-latitude seaway

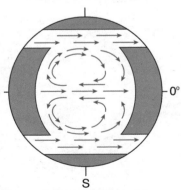

S

3. Double-slice world
with high-latitude seaways

23. Longitudinal continents and ocean circulation

Deep-ocean circulation

Deep-ocean circulation is also an important consideration as it influences the circulation of the surface ocean and the distribution between the hemispheres. The presence or absence of ocean gateways has a profound effect on the deep-ocean circulation. For example, our modern day North Atlantic Deep Water (NADW), which helps to pull the Gulf Stream northwards maintaining the mild European climate may be only 4 million years old. If we run computer simulations of ocean circulation with and without the

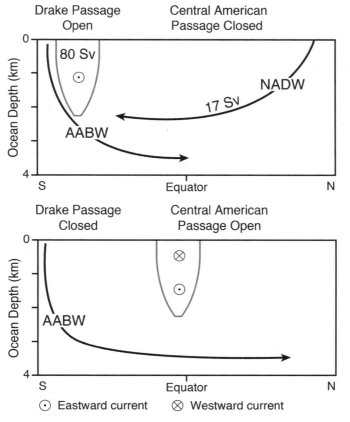

24. Ocean gateways and deep-ocean circulation

Drake Passage and the Panama Gateway, only the modern day combination produces significant NADW. Hence our modern day deep-ocean circulation is due to an open Drake Passage from about 25 million years ago and the closure of the Panama Gateway from about 4 million years later (Figure 24). It is all due to salt. Because of the greater effect of evaporation in the North Atlantic region, the North Atlantic Ocean is saltier than the Pacific Ocean. NADW forms today when the warm, salty water from the Caribbean travels across the Atlantic Ocean and cools down. The high salt load and colder temperature act together to increase the density of the water so it is able to sink north of Iceland. So when the Panama passage way is open then fresher Pacific Ocean water leaks in and reduces the overall salt content of the North Atlantic Ocean. The surface water even when it is cooled is thus not dense enough to sink and so not as much NADW can be formed compared to today. So, one of the fundamental elements of our modern climate system, the competition between the Antarctic Bottom Water and the North Atlantic Deep Water, turns out to be a very young feature.

Vertical tectonics

As the tectonic plates move around the surface of the Earth they frequently clash together, when this happens land is pushed upwards. In some cases chains of mountains are formed or when whole regions are uplifted plateaus are formed. These have a profound effect on the climate system. One of these effects is a rain shadow, which is a dry area on the leeward side of a mountain system. There is usually a corresponding area of increased precipitation on the forward side. As a weather system at ground level moves towards a mountain or plateau it is usually relatively warm and moist (Figure 25). As the air encounters the mountain it is forced to move up and over it. Because of decreasing atmospheric pressure with increasing altitude, the air has to expand and as it does it cools down. Cool air can hold less moisture than warm air so the relative humidity rapidly rises

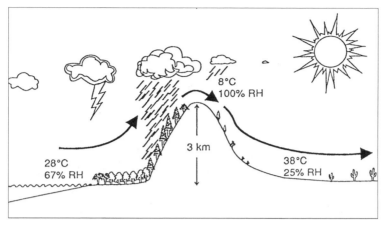

25. Mountain rain-shadow (RH = relative humidity)

until it hits 100 per cent and strong rainfall occurs. As the air descends on the other side of the mountain atmospheric pressure increases and the air temperature rises and the relative humidity drops very low as little or no moisture is left in the air. Hence on the descending side there is a rain shadow as there is no moisture left with which to form rain and this can lead to the creation of a desert. This simple process can control the wetness or dryness of whole continents. Figure 26 shows the effect of whether mountains occur on the western or eastern boundary of a continent. As we saw in Chapter 2 there are three main rainfall belts in the world, one in the tropics and one in the mid-latitudes in each hemisphere. Air in the tropics moves from east to west, while in the mid-latitudes it moves west to east. So having mountains on the western side produces more rainfall on land and produces a wetter continent overall. By coincidence at the moment we have western mountain ranges running down the west coast of North America, the Rockies, and the west coast of South America, the Andes. These mountains not only produce significant wet areas but also famous deserts like the Atacama Desert in Chile and Death Valley in the United States, which are two of the driest deserts on Earth. The contrast between wet

Climate

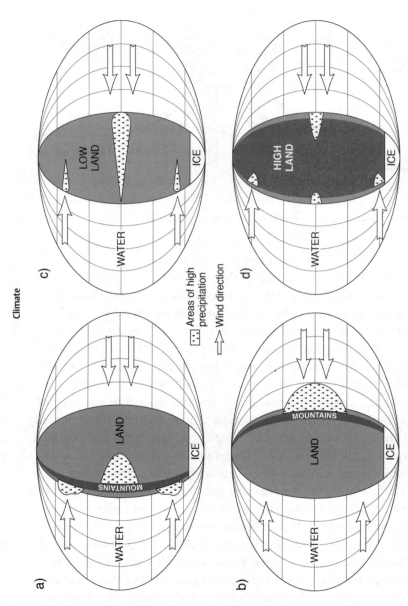

a)

c)

b)

d)

Areas of high precipitation

Wind direction

26. Mountain and plateau effects on global rainfall

70

and dry regions is even sharper if the uplift produces a plateau. Figure 26 shows how little rainfall can make it into a plateau due to this rain shadow effect.

Atmospheric barriers

When huge mountains or plateaus are thrust high up in the sky they interfere with the circulation of the atmosphere. Not only do they force air up and over them but in many cases they deflect the weather system around them. This effect is compounded as uplift areas also warm up in summer and cool down in winter more than the surrounding lowlands. Figure 27 shows that if all the continents in the Northern Hemisphere were flat then the major circulation of the atmosphere would be nearly circular, with maybe a slight deflection due to the difference between land and the oceans. However, if you put the two modern plateaus in place, in other words, the uplifted regions of the Tibetan-Himalayan and Sierran-Coloradan plateaus then there are huge changes in circulation. Both these plateaus are massive. The Tibetan plateau is the world's highest and largest with an area of 2.5 million km^2, which is about four times the size of France. While the Colorado Plateau covers an area of 337,000 km^2 and is joined to numerous other plateaus which make up the Sierran-Coloradan uplift complex.

In Northern Hemisphere summers these two major plateaus heat up more than the surrounding areas and thus the air above them rises creating a low-pressure zone. This sucks in surrounding air creating a cyclonic circulation deflecting weather system much further north and south. In Northern Hemisphere winters these highlands are much colder than the surrounding areas creating a high-pressure system and out-blowing anti-cyclonic circulation (see Figure 27). This deflects Arctic air northwards and keeps the middle of the Asian and North American continents warmer than they would otherwise be. The atmospheric circulation becomes even more complicated when large ice sheets are present on Greenland, North America, and Europe. Because ice sheets are always cold they produce permanent high-pressure systems with out-blowing anti-cyclonic circulation,

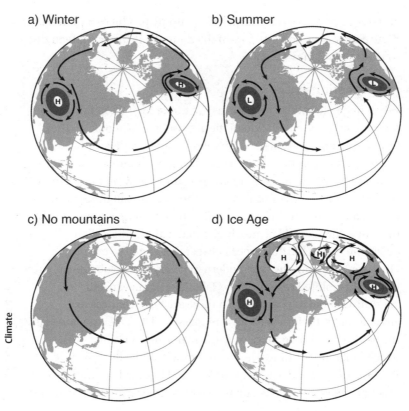

a) Winter   b) Summer

c) No mountains   d) Ice Age

27. **Plateaus and ice sheet effects on atmospheric circulation**

which is discussed in Chapter 7. The summer cyclonic circulation around the Tibetan-Himalayan plateau also creates the Southeast Monsoonal system. Because part of the air that is pulled towards the Himalayas comes from the Indian Ocean it brings with it a lot of moisture. The resultant rainfall is essential for the well-being of two-fifths of the world's population.

Volcanic eruptions

Plate tectonics control the development of volcanoes, which have an important influence on climate through the introduction of gases and dust into the atmosphere. Normal sized volcanoes inject

Climate

sulphur dioxide, carbon dioxide, and dust into the troposphere and can have a considerable effect on our weather. For example in 1883 Krakatoa erupted, killing 36,417 people. The eruption is considered to be the loudest sound ever heard in modern history, with reports of it being heard nearly 3,000 miles away. It was equivalent to 200 megatons of TNT, which is about 13,000 times the nuclear yield of the Little Boy bomb that devastated Hiroshima, Japan, during World War II. The sulphur dioxide and dust injected into the atmosphere increased the amount of sunlight reflected back into space and average global temperatures fell by as much as $1.2\,°C$ in the year following the eruption. Weather patterns continued to be chaotic for years and temperatures did not return to normal until 1888.

On the 15 June 1991 Mount Pinatubo erupted sending 20,000,000 tonnes of sulphur dioxide into the atmosphere. The sulphur dioxide oxidized in the atmosphere to produce a haze of sulfuric acid droplets, which gradually spread throughout the lower stratosphere over the year following the eruption. This time modern instruments were able to measure its effects, which included a 10 per cent reduction in the normal amount of sunlight reaching the Earth's surface. This led to a decrease in Northern Hemisphere average temperatures of 0.5–$0.6\,°C$ and a global decrease in temperature of about $0.4\,°C$.

Both Krakatoa and Pinatubo had a short-term transient effect on climate. This is because the sulphur dioxide and dust were injected relatively low in the atmosphere and the amount of water also injected meant much of the material was washed out of the atmosphere within a few years (Figure 28). However these two eruptions are very small compared to eruptions from supervolcanoes. These are thousands of times larger than Krakatoa. They can occur when magma in the Earth rises into the crust from a hotspot but is unable to break through the crust. Pressure builds in a large and growing magma pool until the crust is unable to contain the pressure. They can also form at convergent

a) Volcanic eruption

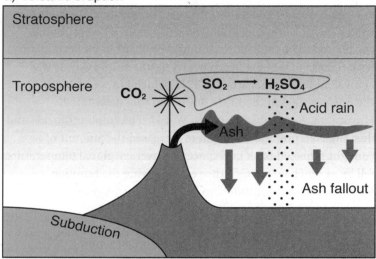

b) Super-volcanic eruption

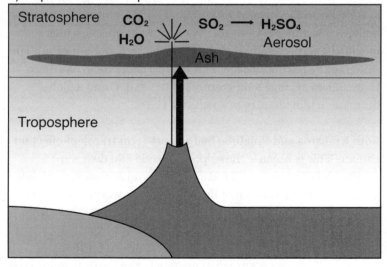

28. **Volcanic eruption effects on atmosphere composition**

plate boundaries, for example Toba, which last erupted about 74,000 years ago and ejected about 2,800 km³ of material into the atmosphere. They can also form in continental hotspot locations, for example Yellowstone, which last erupted 2.1 million years ago and ejected 2,500 km³ of material. Because of the scale of these events the sulphur dioxide and dust are injected much higher in the atmosphere and therefore the effects on the global climate can be much longer. Modelling work by the UK Meteorological Office suggested a tropical supervolcano eruption would cause a drop in global temperatures of at least 6°C, with up to 15°C in the tropics for at least 3 years. Then over a decade the climate would slowly came back to within 1°C of normal. The final effects would take up to a hundred years to get rid of and would be devastating for us if it ever happened. However, in geological terms it is a very short-term event with no significant long-term effect on the climate system.

Icehouse and greenhouse worlds

Plate tectonics drives the slow shift of the continents across the globe, shifting from a supercontinent to fragmented continents and then back again. The supercontinent Rodinia formed about 1.1 billion years ago and broke up roughly 750 million years ago. One of the fragments included large parts of the continents we now find in the Southern Hemisphere. Plate tectonics brought the fragments of Rodinia back together in a different configuration about 300 million years ago, forming the best-known supercontinent, Pangaea. Pangaea subsequently broke up into the northern and southern supercontinents of Laurasia and Gondwana, about 200 million years ago. Both of these supercontinents have continued to fragment over the last 100 million years. Icehouse climates form when the continents are moving together. The sea level is low due to lack of seafloor production. The climate becomes cooler and more arid, because of the reduction in rainfall due to the strong rain shadow effect of large superplateaus. Greenhouse climates, on the other hand, are formed as the continents disperse, with sea levels high due to the high level of sea floor spreading. There are relatively high levels of carbon dioxide in the atmosphere, possibly over three times the current levels, due to production at oceanic rifting zones. This produces a warm and humid climate.

The formation and break up of these supercontinents had a huge effect on evolution. Supercontinents are extremely bad for life. First, there is a massive reduction in the amount of shelf sea areas, where we think multi-cellular life may have started. Second, the interior of continents are very dry and global climate is usually cold. A number of key mass extinctions are correlated with the formation of supercontinents. For example it is estimated that up to 96 per cent of all marine species and 70 per cent of terrestrial vertebrate species became extinct during the Permian–Triassic extinction event 250 million years ago, which is nicknamed the 'mother of all mass extinctions' (Figure 29). It is also not surprising the explosion of complex, multi-cellular organisms occurred during the Cambrian period about 550 million years ago, following the break up of the Rodinia supercontinent.

Snowball Earth

Prior to about 650 million years ago there is an idea that the surface of the Earth became entirely frozen at least once—the so-called Snowball Earth hypothesis. It is a way to explain the sedimentary deposits found in the tropics, which show glacial features that suggest there must have been a lot of ice in the tropics. Opponents of the idea suggest that the geological evidence does not suggest a global freezing. Moreover there is difficulty in getting the whole ocean to become ice- or even slush-covered. There is also the difficulty of seeing how the world, once in a snowball state, would subsequently escape the frozen condition. One answer is that this would occur through the slow build up of atmospheric carbon dioxide and methane, which would eventually reach a critical concentration, warming the atmosphere enough to start the melting process. There are a number of unanswered questions, including whether the Earth was a *full* snowball or a 'slushball' with a thin equatorial band of open water. But what is particularly

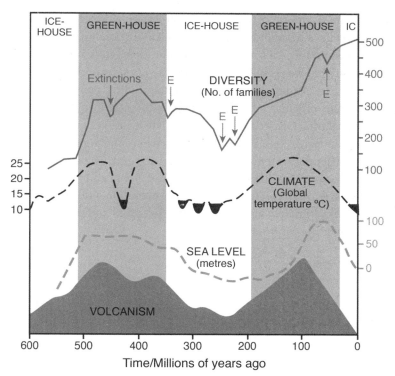

29. Long-term links between tectonics, sea level, climate, biodiversity, and extinctions

interesting is the idea that the evolution of complex life put an end to the possibilities of ever having a snowball Earth again. Professor Andy Ridgwell at Bristol University has suggested that the evolution of marine mirco-organisms that form calcite shells now buffers the oceans' carbonate system so much that the extreme variation in atmospheric carbon dioxide needed to plunge the world into or out of a snowball or slushball condition could not now occur.

Summary

Our modern climate system is a product of the slow movements of the continents across the face of the Earth. We are currently in an 'icehouse world', as we have continents on or surrounding each pole. The reduction of atmospheric carbon dioxide has allowed the growth of permanent ice sheets on Antarctica and Greenland. This has produced a very strong Equator–pole temperature gradient of at least $60\,°C$, which drives a very vigorous climate system. The current arrangements of longitudinal continents and ocean gateways, has produced strong deep-water formation in the North Atlantic Ocean and Antarctica. The location of modern mountain ranges and plateaus controls where the major deserts and monsoonal systems of the world are located. The movement of continents has also profoundly affected global and regional climates, which have in turn influenced evolution. Our modern climate is ultimately a product of plate tectonics and the random location of the continents.

Chapter 6
Global climate cooling

Introduction

Fifty million years ago the Earth was a very different place. The world was both warmer and wetter, with rainforest extending all the way up to northern Canada and all the way down to Patagonia. So how did we go from this lush, vibrant Earth to the ice-locked, cool planet we have today. What caused the beginning of the great ice ages? If you compare a map of the world 50 million years ago with one of the world today they seem to be the same, until you look in detail. We saw in Chapter 5 that movements of the continents around the face of the planet are very slow, but minor changes in location have had a profound effect on global climate. Over the last 50 million years these small changes have moved the Earth's climate from a being greenhouse to an icehouse world.

The last 100 million years

For the last 100 million years Antarctica has sat over the South Pole and the Americas and Asian continent have surrounded the North Pole. But only for the last 2.5 million years have we cycled in and out of the great ice ages, the so-called glacial–interglacial cycles. There must, therefore, be additional factors controlling the temperature of the Earth. In particular you need a means of cooling down the continents on or surrounding a pole. In the case of Antarctica the ice did not start building up until about 35 million years ago (Figure 30). Prior to that Antarctica was covered by lush, temperate forest: bones of dinosaurs have been found there dating from before they went extinct 65 million years ago. What changed 35 million years ago was a culmination of minor tectonic movements. Slowly South America and Australia are moving away from Antarctic. About 35 million years ago the ocean opened up between Tasmania and Antarctica. This was followed about 30 million years ago by the opening of the Drake Passage between South America and Antarctica, one of the most feared stretches of ocean. This allowed the Southern Ocean to start circulating around Antarctica. The Southern Ocean acts very much like the fluid in your refrigerator at home. It takes heat from Antarctica as it flows around the continent and then releases it into the Atlantic, Indian, and Pacific Oceans, into which it mixes. Opening up these seemingly small ocean gateways between the continents produced an ocean that can circulate around Antarctica completely, continually sucking out heat from the continent. So efficient is this process that there is now enough ice on Antarctica that if all of it melted the global sea level would rise over 65 metres—high enough to cover the head of the Statue of Liberty. This tectonic cause of the glaciation of Antarctic is also the reason that scientists are confident that global warming will not cause the eastern Antarctic ice sheet melt—if it were to melt it would cause an approximate 60 metre rise in sea level. The same cannot be said of the unstable western Antarctic ice sheet (see Chapter 8).

The ice-locked Antarctica of 30 million years ago did not, however, last long. Between 25 and 10 million years ago Antarctica ceased to be completely covered with ice. The question is why did the world start to cool all over again 10 million years ago and why did the ice start building up in the Northern Hemisphere? Palaeoclimatologists believe that relatively low levels of atmospheric carbon dioxide are essential if you are to maintain a cold planet. Computer models have shown that if you have high levels of atmospheric carbon

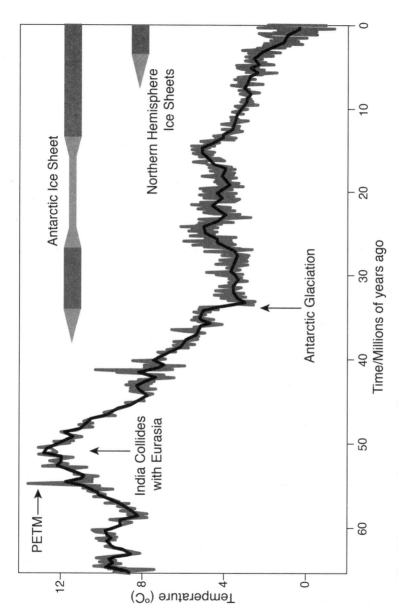

30. Global climate over the last 65 million years

81

dioxide you cannot get ice to grow on Antarctica even with the ocean heat extractor. So what caused the carbon dioxide to get lower and why did the ice start growing in the north?

What caused the big freeze?

In 1988 Professor Bill Ruddiman and his then graduate student Maureen Raymo while at the Lamont-Doherty Earth Observatory wrote an extremely influential paper. They suggested that global cooling and the build up of ice sheets in the Northern Hemisphere were caused by uplift of the Tibetan-Himalayan and Sierran-Coloradan regions. As we saw in Chapter 5 huge plateaus can alter the circulation of the atmosphere and they argued this cooled the Northern Hemisphere, allowing snow and ice to build up. However, what they did not realize at the time was most of the Himalayan uplift occurred much earlier between 20 and 17 million years ago and thus it was too early to have been the direct cause of the ice in the north. But Maureen Raymo then came up with a startling suggestion that this uplift may have caused a massive increase in erosion that uses up atmospheric carbon dioxide in the process. This is because when you make a mountain range you also produce a rain shadow. So, one side of the mountain has a lot more rain on it as the air is forced up and over the mountain. This is also why mountains erode much faster than gentle rolling hills. She argued that this extra rainwater and carbon dioxide from the atmosphere form a weak carbonic acid solution, which dissolves rocks. But interestingly only the weathering of silicate minerals makes a difference to atmospheric carbon dioxide levels, as weathering of carbonate rocks by carbonic acid returns carbon dioxide to the atmosphere. As much of the Himalayas is made up of silica rocks there was a lot of rock that could lock up atmospheric carbon dioxide. The new minerals dissolved in the rainwater are then washed into the oceans and used by marine plankton to make shells out of the calcium carbonate. The calcite skeletal remains of the marine biota are ultimately deposited as deep sea sediments and hence lost from the global carbon cycle for the duration of the lifecycle of the oceanic crust on which they have been deposited. It's a fast track way of getting atmospheric carbon dioxide out of the atmosphere and dumping it at the bottom of the ocean. Geological evidence for long-term changes in atmospheric carbon dioxide does support the idea that it has dropped significantly over the last 20 million years. The only problem scientists have with this theory is what stops this process. With the amount of rock in Tibet that has been eroded over the last 20 million years all the carbon dioxide in the atmosphere should have been stripped out. So there must be other natural mechanisms which help to maintain the balance of carbon dioxide in the atmosphere as the long-term concentration of carbon dioxide in the atmosphere is the result of a balance between what is removed by weathering and deposition in the deep ocean and the amount recycled by subduction zones and emitted by volcanoes.

With atmospheric carbon dioxide levels dropping between 10 and 5 million years ago the Greenland ice sheet started to build up. Interestingly Greenland started to glaciate from the south first. This is because you must have a moisture source to build ice with. So by 5 million years ago we had huge ice sheets on Antarctica and Greenland, very much like today. The great ice ages when huge ice sheets waxed and waned on North America and Northern Europe did not start until 2.5 million years ago, however there is intriguing evidence suggesting that around 6 million years ago these big ice sheets did start to grow. Rock fragments from the continent, eroded by ice and then dumped at sea by icebergs have been found in the North Atlantic Ocean, North Pacific Ocean, and Norwegian Sea at this time. This seems to have been a failed attempt to start the great ice ages and could be because of the Mediterranean Sea.

The great salt crisis

About 6 million years ago the gradual tectonic changes resulted in the closure of the Strait of Gibraltar. This led to the transient isolation of the Mediterranean Sea from the Atlantic Ocean. During this isolation the Mediterranean Sea dried out several times, creating vast evaporite (salt) deposits. Just image a huge version of the Dead Sea where a few metres of seawater cover a vast area. This event is called the Messinian Salinity Crisis and it was a global climate event because nearly 6 per cent of all dissolved salts in the world's oceans were removed. By 5.5 million years ago the Mediterranean Sea was completely isolated and was a salt desert (Figure 31). This was roughly the same time as palaeoclimate records indicate that the Northern Hemisphere was starting to glaciate. But at about 5.3 million years ago the Strait of Gibraltar reopened, causing the Terminal Messinian Flood, also known as the Zanclean Flood or Zanclean Deluge. Scientists have envisaged an immense waterfall higher than today's Angel Falls in Venezuela (979 m), and far more powerful than either the Iguazu Falls on the boundary between Argentina and Brazil or the Niagara Falls on the boundary between Canada and the USA. More recent studies of the underground structures at the Gibraltar Strait show that the flooding channel may have descended in a rather more gradual way to the dry Mediterranean. The flood could have occurred over months or a couple of years, but it meant that large quantities of dissolved salt were pumped back into the world's oceans via the Mediterranean–Atlantic gateway. This stopped the Great Ice Age in its tracks, and was entirely due to how oceans circulate. As we saw in Chapter 2 the Gulf Stream not only keeps Europe warm but also drives the deep-ocean circulation and keeps the whole planet relatively warm. Five million years ago the deep-ocean circulation was not as strong as it is today. This is because fresher Pacific Ocean water was still able to leak through the Panama ocean gateway which is discussed below. So the sudden massive increase in salt due to the Terminal Messinian Flood increased the salt in the North Atlantic Ocean ensuring a very vigorous Gulf Stream and sinking water in the Nordic Seas. With all this tropical heat being efficiently pumped northwards the slide into any further great ice ages was halted about 5 million years ago. We had to wait another 2.5 million years before the global climate was ready to try again.

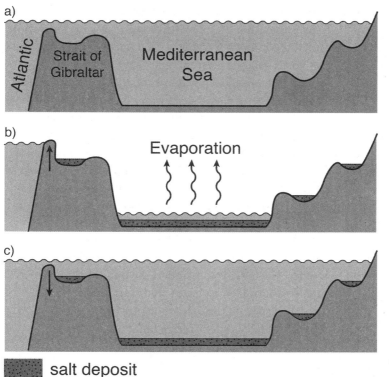

a)

Atlantic

Strait of Gibraltar

Mediterranean Sea

b)

Evaporation

c)

Global climate cooling

salt deposit

31. Mediterranean Messinian 'Salinity Crisis' and 'Terminal Flood' approximately 5 million years ago

The Panama paradox

Another important tectonic control, which geologists believe to be a trigger for causing great ice ages is the closure of the Pacific–Caribbean gateway. Professors Gerald Haug and Ralf Tiedemann now at Zurich University and the Alfred Wegener Institute used evidence from ocean sediments to suggest that the Panama ocean gateway began to close 4.5 million years ago and finally closed about 2 million years ago. The closure of the Panama gateway, however, causes a paradox, as it would have both helped and hindered the start of the Great Ice Age. First, the reduced inflow of Pacific surface water to the Caribbean would have increased the salinity of the Caribbean, because the Pacific Ocean water is fresher than that of the North Atlantic Ocean. This would have increased the salinity of water carried northward by the Gulf Stream and North Atlantic Current, and as we have seen above this would have enhanced deep water formation. The increased strength of the Gulf Stream and deep water formation would have worked against the start of the Great Ice Age as it enhances the oceanic heat transport to the high latitudes, which would have worked against ice sheet formation. So after the aborted attempt to start the Great Ice Age about 5 million years ago the progressive closure of the Panama ocean gateway continued to increase the heat transport northward keeping the chill at bay. But here is the paradox: two things are needed to build large ice sheets—cold temperatures and lots of moisture. The enhanced Gulf Stream also pumped much more moisture northward, ready to stimulate the formation of ice sheets. This meant that the building of large ice sheets in the Northern Hemisphere could start at a warmer temperature because of all the extra moisture being pumped northward ready to fall as snow and to build up ice sheets.

Why 2.5 million years ago?

Tectonic forcing alone cannot explain the amazingly fast intensification of Northern Hemisphere glaciation (Figure 32). Work of mine using ocean sediments suggests that there were three main steps in the transition to the Great Ice Ages. The evidence is based on when rock fragments that had been ripped off the continent by ice were deposited in the adjacent ocean basin by icebergs. First, sheets started growing in the Eurasian Arctic and Northeast Asia regions approximately 2.74 million years ago, with some evidence of growth of the Northeast American ice sheet; second, an ice sheet started to build up on Alaska 2.70 million years ago; and, third, the biggest ice sheet of them all, on the Northeast American continent, reached its maximum size 2.54 million years ago. So in less than 200,000 years we go from the warm, balmy conditions of the early Pliocene, which Professor

Michael Sarnthein of Kiel University called 'the golden age of climate', to the Great Ice Ages.

The timing of the start of the intensification of Northern Hemisphere glaciation must have had another cause. It has been suggested that changes in orbital forcing (changes in the way the

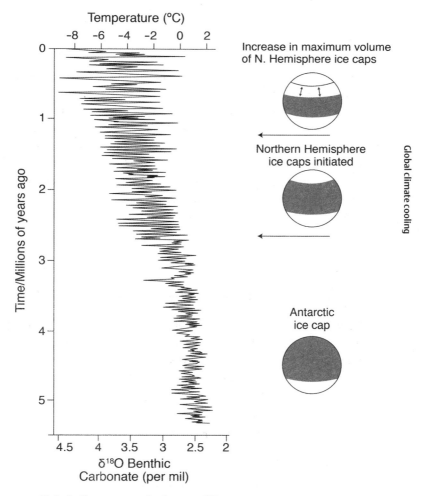

32. Global climate over the last 5 million years

Earth spins round the sun) may have been an important mechanism contributing to the global cooling. The details of the Earth's numerous wobbles and how they caused the waxing and waning of individual ice ages are discussed in the next chapter. But though these individual wobbles are on the scale of tens of thousands of years, there are much longer variations as well. For example one of the most important is obliquity or tilt, which is the wobble of the Earth's axis of rotation up and down—or, put another way, the tilt of the Earth's axis of rotation with respect to the plane of its orbit. Over a period of 41,000 years the Earth's axis of rotation will lean a little more towards the sun and then a little less. It's not a large change, varying from between 21.8° and 24.4°. In Chapter 1 we explored how the tilt of the axis of rotation gives us our seasons. Hence a larger tilt will result in a larger difference between summer and winter. Over a period of 1.25 million years the amplitude of the tilt changes. Both times the Earth tried to glaciate the Northern Hemisphere, at 5 million years and 2.5 million years ago, the variation of tilt increased to its largest value. This made the changes in each season very marked, most important were the cold summers in the north ensuring that the ice did not melt and could develop into ice sheets.

The tropics react to the ice ages

The onset of the intensification of Northern Hemisphere glaciation did not just affect the high latitudes. It seems half a million years after the start of the Great Ice Ages things changed in the tropics. Before 2 million years it seems there was a very slight east–west sea-surface temperature gradient in the Pacific Ocean, but this gradient later grew, showing a switch in the tropics and sub-tropics to a modern mode of circulation with relatively strong Walker Circulation and cool sub-tropical temperatures. The Walker Circulation is the atmospheric east–west component of the Hadley Cell and is instrumental in controlling rainfall in the tropics. The Walker Circulation is also a key element in the El Niño–Southern Oscillation (ENSO; see Chapter 3). So before 2 million years ago ENSO may not have existed in its modern form because there was a relatively weak Walker Circulation. The development of the Walker Circulation also seems to be linked to early human evolution. The strengthening of the east–west circulation seems to have produced deep, fresh but ephemeral lakes in the East African Rift. It has recently been postulated that this distinctive climate pulse with rapid appearance and disappearance of lakes at about 2 million years may be linked with the evolution of *Homo erectus* in Africa with an over 80 per cent increase in brain size and our ancestors migrating out of Africa for the very first time.

Mid-Pleistocene Transition

Mid-Pleistocene Transition (MPT) is the marked prolongation and intensification of glacial–interglacial climate cycles that occurred sometime after 800,000 years ago (Figure 32). Prior to the MPT, the glacial–interglacial cycles seem to occur every 41,000 years, which corresponds to the slow changes in the tilt of the Earth. After about 800,000 years ago the glacial–interglacial cycles seem to be much longer, averaging over 100,000 years. The shape of these cycles also changes. Before the MPT the transition between glacial and interglacial periods is smooth and the world seemed to spend about equal time in each climate. After the MPT the cycles became saw-toothed with ice building up over 80,000 years to produce deep, intense ice ages and then rapid deglaciation, with the loss of all that ice within 4,000 years. The climate then stayed in an interglacial period resembling our current climate for about 10,000 years before descending back into an ice age. One suggestion for this saw-toothed pattern is that the much larger ice sheets are very unstable and therefore with a slight change in climate they collapsed rapidly and the whole climate system rebounded back into an interglacial period. In the next chapter we will examine these recent glacial–interglacial cycles in more detail.

Chapter 7
Great ice ages

Introduction

In 1658 Archbishop Ussher of Armagh looked at the features of the landscape around him and attributed them to Noah's Flood. Using the Bible he diligently dated the flood and thus the landscape to 4004 BC. It was not until 1787 that Horace-Bénédict de Saussure, a Genevan aristocrat, physicist, and Alpine traveller, recognized that Alpine erratic boulders had been moved hundreds of miles down the slopes of the Jura Range and reasoned that the mountain glaciers must have extended much further into the past. This discovery had to wait until 1837 for Louis Agassiz, the Swiss geologist, to put forward his 'ice age' or 'glacial' theory based on the evidence of erratic boulders and end moraines. Terminal moraines are hills formed by eroded sediment being pushed in front of an advancing ice sheet like a bulldozer. When the ice sheet reaches its maximum extent the sediment is deposited as a line of hills which trace the front edge of the ice sheet. In 1909 Albrecht Penck and Eduard Brückner, German and Austrian geographers, produced a 3-volume work entitled *Die Alpen im Eiszeitalter* (*The Alps in the Ice Age*). They concluded that there were four major 'ice age' or glacial periods, the Gunz, Mindel, Riss, and Wrum. Terrestrial or land based evidence has the disadvantage of being discontinuous and the evidence may be destroyed by subsequent ice sheet advances. Hence it was not until the 1960s when long, continuous sediment cores were recovered from the bottom of the oceans that it was realized how many ice ages there had been. We now have the ability to drill in the ocean to a depth of four miles and still be able to recover over half a mile of sediment below the sea floor. From studying these marine sediments scientists have documented 50 ice ages that occurred in the last 2.5 million years.

Waxing and waning of the great ice ages

We now know that glacial–interglacial cycles are the fundamental characteristic of the Quaternary Period, the last 2.5 million years. The waxing and waning of the huge continental ice sheet is initiated by the changes in the Earth's orbit around the sun. The Earth over long periods of time wobbles on its axis and thus changes the amount of sunlight or solar energy received by different parts of the Earth. These small changes are enough to push or force climate change. However, these waxings and wanings are not caused by Earth's orbital wobbles, but rather by the Earth's climate reaction, which translates relatively small changes in regional solar energy into major climatic variability. For example, the position of the Earth is very similar today to what it was 21,000 years ago during the last ice age. So it is not the exact orbital position that controls the climate but rather the changes in the orbital positions. There are three main orbital parameters or wobbles, called eccentricity, obliquity or tilt, and precession (see Box 4), and as you can see from the box each has a unique cycle and effect on climate. More exciting though is when we combine them all together and see how they push the climate either into or out of a great ice age.

Clockwork climate?

Combining the effects of all three orbital parameters you can calculate the solar energy received for any latitude back through time. Milutin Milankovitch, a brilliant Serbian mathematician and climatologist, in 1949 suggested that summer insolation at

Box 4 Orbital forcing

There are three main orbital parameters or wobbles: eccentricity, obliquity (tilt), and precession (Figure 33), which have a significant effect on the long-term climate of the Earth.

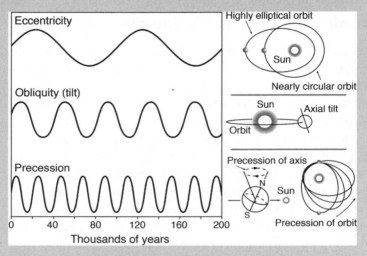

33. Orbital variables

Eccentricity is the shape of the Earth's orbit around the sun, which varies from a circle (precession) to an ellipse. These variations occur over a period of about 96,000 years with an additional long cycle of about 400,000 years. Described another way, the long axis of the ellipse varies in length over time. In recent times, the Earth is closest to the sun on 3 January at about 146 million km; this position is known as perihelion. On 4 July the Earth is at its greatest distance from the sun, at about 156 million km; this position is known as the aphelion. Changes in eccentricity cause only very minor variations in total insolation, but can have significant seasonal effects when combined with precession.

If the orbit of the Earth were perfectly circular there would be no seasonal variation in solar insolation. Today, the average amount of radiation received by the Earth at perihelion is ~351 W/m² and is 329 W/m² at aphelion. This represents a difference of ~6 per cent, but at times of maximum eccentricity (ellipse length) over the last 5 million years the difference could have been as great as 30 per cent. Milutin Milankovitch in 1949 suggested that northern ice sheets are more likely to form when the sun is more distant in summer, so that each year some of the previous winter's snow survives. The other effect of eccentricity is to modulate the effects of precession. However, it is essential to note that eccentricity is by far the weakest of all three orbital parameters.

Obliquity or the tilt of the Earth's axis of rotation with respect to the plane of its orbit varies between 21.8° and 24.4° over a period of 41,000 years. It is the tilt of the axis of rotation that causes the seasons, as described in Chapter 1. The larger the obliquity, the greater the difference between the insolation received in summer and winter. Milutin Milankovitch suggested that the colder the Northern Hemisphere summers, the more likely it is that snowfall will accumulate, resulting in the gradual build up of glaciers and ice sheets.

Precession has components that relate to the elliptical orbit of the Earth (eccentricity) and its axis of rotation. The Earth's rotational axis precesses every 27,000 years. This is similar to the gyrations of the rotational axis of a toy spinning top. Precession causes a change in the Earth–sun distance for any particular date, for example, the beginning of the Northern Hemisphere summer. It is the combination of the different orbital parameters that lead to the two different precessional periodicities of 23,000 years and 19,000 years. Combining the precession of the axis of rotation plus the precessional changes in orbit produces a period of 23,000 years. However, the combination of eccentricity (96,000 years) and precession of the axis of rotation also results in a period of 19,000 years. These two periodicities combine so that perihelion coincides with the summer season in each hemisphere on average every 21.7 thousand years. Precession has the most significant impact in the tropics (in contrast to the impact of obliquity at the Equator, which is zero). So although obliquity clearly influences high latitude climate change, which may ultimately influence the tropics, the direct effects of insolation in the tropics are due to eccentricity-modulated precession alone.

Combining the effects of eccentricity, obliquity, and precession provides the means for calculating insolation for any latitude through time. Figure 34 shows the calculated insolation for 65°N compared with the changing size of the ice sheets represented by global sea level change for the last 600,000 years.

65°N which is just south of the Arctic Circle was critical in controlling glacial–interglacial cycles (Figure 34). He argued that if the summer insolation was reduced enough then ice could survive through the summer, start to build up, and eventually produce an ice sheet. Orbital forcing does have a large influence on this summer insolation; the maximum change in solar radiation in the last half million years is equivalent to reducing the amount of summer radiation received today at 65°N to that received now over 550 km to the north at 77°N. In simplistic terms, this brings the current ice limit in mid-Norway down to the latitude of mid-Scotland. These lows in 65°N insolation are caused by eccentricity elongating the summer Earth–sun distance, obliquity being shallow and precession placing the summer season at the longest Earth–sun distance produced by eccentricity. The reason why it is at 65°N and not at 65°S, which controls climate, is very simple. Any ice that builds up in the Northern Hemisphere has many continents to grow upon. In contrast in the Southern Hemisphere the ice growth is limited by the Southern Ocean, as

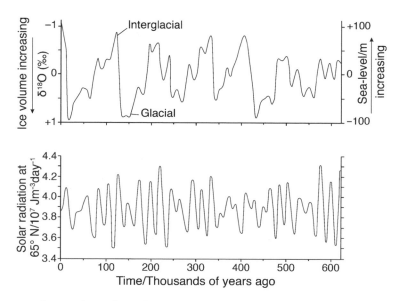

34. Comparison of Northern Hemisphere solar radiation and global sea level

any extra ice produced on Antarctica falls into the ocean and is swept away to warmer seas (Figure 35). So the conventional view of glaciation is that low summer solar energy in the temperate Northern Hemisphere allows ice to survive summer and thus ice sheets start to build up on the northern continents. But this apparently simple clockwork view of the world is really much more complicated, as the effects of orbital changes on the seasons is very small and it is feedback in the climate system that amplifies these changes.

What causes the glacial–interglacial cycles

Orbital forcing in itself is insufficient to drive the observed glacial–interglacial variability in climate. Instead, the Earth system amplifies and transforms the changes in solar energy received at the Earth's surface through various feedback mechanisms. For example, let us start with building an ice age.

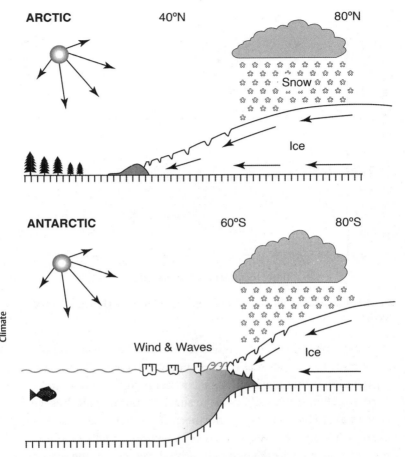

ARCTIC   40°N   80°N

Snow

Ice

ANTARCTIC   60°S   80°S

Wind & Waves   Ice

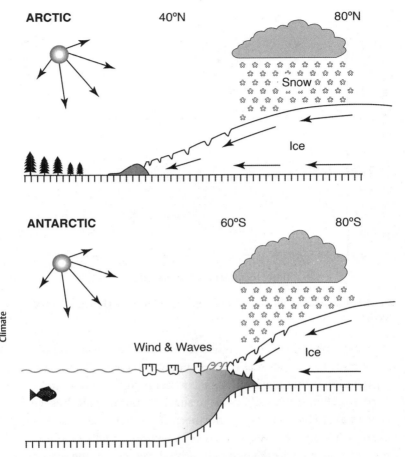

35. Ice sheet expansion in the Arctic and Antarctic

The first thing that needs to happen is a slight reduction in summer temperatures. As snow and ice accumulate due to this change in summer temperature, the albedo, the reflection of sunlight back into space, increases. The process of reflecting more sunlight back into space suppresses local temperatures, which promotes the accumulation of more snow and ice, which further increases the albedo of the region, producing the so-called 'ice–albedo' feedback. So once you have a small ice sheet it changes the environment around it to make more snow and ice, and will get bigger and bigger.

Another feedback is triggered when the ice sheets, particularly the Laurentide ice sheet on North America, become big enough to deflect the atmospheric planetary waves (see Chapter 5, Figure 27). This changes the storm path across the North Atlantic Ocean and prevents the Gulf Stream and North Atlantic Drift from penetrating as far north as it does today. This surface ocean change combined with the general increase in melt-water in the Nordic Seas and Atlantic Ocean due to the presence of a large continental ice sheet ultimately leads to a reduction in the production of deep water. Deep-water production in the Greenland and Labrador Seas is the heartbeat of our modern climate. By reducing the formation of deep water it reduces the amount of warm water pulled northwards, all of which leads to increased cooling in the Northern Hemisphere and expansion of the ice sheets.

There is currently a debate amongst palaeoclimatologists about the role of the physical climate feedbacks described above and the role of greenhouse gases in the atmosphere. Air bubbles trapped in polar ice have shown us that carbon dioxide dropped by a third and methane by a half during each glacial period. These changes would have compounded the cooling that occurred during each glacial period, helping to build more ice. So the debate continues: do changes in the Earth's orbit affect the production of greenhouse gases, cooling down the Earth to make the Northern Hemisphere continents susceptible to the build up of large ice sheets? Or do changes in the Earth's orbit start the build up of large ice sheets in the Northern Hemisphere that then change global climate and reduce the production of greenhouse gases, thus prolonging and deepening the ice age? I am sorry to say the jury is still out on these questions. However, what we do know is that greenhouse gases played a critical role in glacial—interglacial cycles. We also know that changes in greenhouse gases concentration always come before changes in global temperatures.

An important question is why these feedbacks don't end up running away and freezing the whole Earth. The answer is that they are prevented from having a runaway effect by a process called 'moisture limitation'. To build an ice sheet you need it to be cold and wet. However, as the warm surface water is forced further and further south, supply of the moisture that is required to build ice sheets decreases. So by changing the atmospheric and ocean circulation, the ice sheets end up starving themselves of moisture.

In the last million years it took up to 80,000 years to build up ice sheets to reach the maximum extent of ice. So the last time this occurred was about 21,000 years ago. However, getting rid of the ice is much quicker. This process known as 'deglaciation' usually takes a maximum of only 4,000 years. This deglaciation is triggered by an increase in solar energy received during the summer at about 65°N. This encourages the Northern Hemisphere ice sheets to melt slightly. The rise of atmospheric carbon dioxide and methane promotes warming globally and encourages the melting of the large continental ice sheets. But these processes have to work against the ice sheets' albedo effect, which produces a microclimate that works to keep them intact. What causes the rapid removal of ice is the rise in sea level due to the melting ice sheet, where large ice sheets adjacent to the oceans are undercut by a rising sea level. The coldest sea water can become is about $-1.8\,°C$ while the base of the ice sheet is usually colder than $-30\,°C$. The effect of the water against the ice is similar to putting hot water under a tub of ice cream. Undercutting of the ice sheet leads to more melting and ice caving into the ocean. This in turns increases sea level, which causes even more undercutting. This sea level feedback process can occur extremely rapidly. Once the ice sheets are in full retreat then the other feedback mechanisms discussed above are thrown into reverse.

Anatomy of the last ice age

If we focus on the last ice age, only 21,000 years ago, we can see what a considerable difference glaciation made to the climate of Earth. In North America there was nearly continuous ice spreading across the continent from the Pacific to the Atlantic Ocean. It was made up of two separate ice sheets, the Laurentide ice sheet in the east, centred on Hudson Bay, and the Cordilleran ice sheet in the west, in the coastal ranges and the Rockies. The Laurentide covered over 13 million km² of land and reached over 3,300 m in thickness at its deepest over the Hudson Bay. Its maximum reach extended from New York, to Cincinnati, to St Louis, and to Calgary. In Europe there were two major ice sheets, the Fenno-Scandinavian and the British ice sheets, with a minor one over the European Alps. The British ice sheet during many glacial periods merged with the Scandinavian ice sheet. On average it covered about 340,000 km² per glacial period. During the last glacial period the ice sheet reached halfway down the British Isles to the edge of Norfolk. The Fenno-Scandinavian ice sheet was much larger than the British ice sheet, covering an area of 6.6 million km², and extended all the way from Norway to the Ural Mountains in Russia. We must also not forget the Southern Hemisphere as there were significant ice sheets in Patagonia, South Africa, southern Australia, and New Zealand. In addition the Antarctic ice sheet expanded by about 10 per cent and seasonal sea ice extended an additional 500 miles away from the continent. It is difficult to imagine the huge amount of water that was locked up in these ice sheets. One way of understanding is to consider the oceans. The oceans cover over 70 per cent of our planet and so much water was sucked out of them and locked up in the ice sheets that the sea level dropped by over 120 m. This is approximately the equivalent of the height of the London Eye. If all the ice on Antarctica and Greenland today melted it would raise the sea level by 70 m. During a glacial period global temperatures were 5–6°C lower than today, but this temperature drop was not evenly distributed, with the high latitudes cooling by as much as 12°C, while even the tropics cooled by between 2°C and 5°C. The ice ages were also very dry periods with large amounts of dust in the atmosphere. For example, in Northern China, eastern USA, central and eastern Europe, central Asia, and Patagonia there are deposits of hundreds of metres of dust called 'loess deposits', which built up during the ice age.

Ice shapes the land

During the glacial periods the presence of these huge ice sheets profoundly affected the local climate, vegetation, and landscape. The great boreal forests of the high latitudes were devastated as the land they once occupied was overrun by expanding ice sheets. The reduction of atmospheric moisture greatly reduced rainfall and the great wetlands and tropical rainforest of the world shrank. The huge continental ice sheets also had a profound effect on our landscape. There are very few places in the temperate latitudes that have not been affected by the ice ages. If you travel through northern Europe and North America you will see the dramatic effects of ice ages on the landscape. These effects have made great backdrops for movies, for example the *Lord of the Rings* trilogy was filmed in New Zealand and the mountainous, wild, extreme landscape you see is the result of an ice sheet grinding away over the islands for thousands of years. So, next time you see those movies, think 'ice sheets'. The ice sheets have left us with a legacy of U-shaped valleys, fiords, moraines, and egg shaped hills called 'drumlins'. Even the current position of the River Thames is due to the ice. Previously the River Thames ran through St Albans to the north of London and met the North Sea in Essex. The last but one ice age was so intense in Europe that the ice sheet made it down as far as north London. This re-routed the River Thames to its current path. So the geography of London was primarily controlled by the ice age. In the USA the paths of many major rivers were altered both by the location of an ice sheet and also by the huge amount of meltwater which burst from them as they later melted 12,000 years ago. The pathways of the Laurentide and Mississippi rivers are relics of these great floods at the end of the last glacial period.

The geography of the Earth was also altered as the lowering of global sea level by 120 m meant that the continents changed shape. Islands such as Britain became part of the mainland, meaning that during the last ice age it would have been possible to walk across the English Channel to France. The only thing that would stop you would be the huge new river running down the centre of what is now the English Channel taking water from the Thames, the Rhine, and the Seine out to the Atlantic Ocean. All around the world land bridges were formed by the lowering sea level allowing new species to invade new areas. Around the world, islands such as Sri Lanka, Japan, Britain, Sicily, Papua New Guinea, and the Falklands became part of the adjacent mainland. For example, the chain of islands across the Bering Sea, which separates Northeast Asia from Alaska became joined. So during the end of last glacial period, as the climate started to warm up, humans were able to cross from Asia into North America for the very first time, colonizing the New World.

The case of the missing grass in Amazonia

Ice ages clearly affect the whole of the global climate system, however there is controversy concerning what effects they had on the tropics. Half the surface of the planet lies between the Tropics of Cancer and Capricorn and includes all the tropical rainforest of the world. Of these the most important region in terms of size and species' diversity is the Amazon. The Amazon Basin is the largest in the world covering an area of 7 million km² and it discharges approximately 20 per cent of all freshwater carried to the oceans. The majority of it is covered by extremely diverse rainforest. In 1969 Haffer put forward a wonderful theory linking the ice age to why the Amazon is so diverse. During each glacial period, he suggested, lower temperatures and precipitation in the tropics allowed savannah to replace the majority of the tropical rainforest. However some of the tropical rainforest would have survived in small 'refugia', isolated islands of rainforest surrounded by grassland. These isolated patches of rainforest would have become hotbeds of evolution, producing many new species. At the end of each glacial period the patchwork of rainforest merged back together with higher levels of species diversity and endemism than previously. However, by the late 1990s this theory came under attack, as more and more scientists failed to find the huge increase in savannah. We now know from pollen records and computer models that in the Amazon the combination of dry and cold conditions meant that savannah did encroach a little bit at the edges, reducing the area of the Amazon rainforest to 80 per cent of today's coverage. But it is a testament to the resilience and importance of tropical rainforest in the global ecosystem that the Amazon survived and even flourished during an ice age. One of the reasons why the rainforest survived the glacial period was because the cold conditions actually helped to reduce the problem of less rainfall: the cold temperatures reduced the amount of evaporation from the trees and thus the loss of moisture which is essential for the rainforest. However, there were major changes in the species composition of the Amazon rainforest during glacial periods. For example, we know from pollen records that many of the tree species now found in the Andes were in the heart of the Amazon forest. This is because the species which are more cold-adapted are pushed up to higher 'colder' altitudes during warm interglacial periods. This is important as it means we cannot see the current Amazon rainforest as the normal condition, because for the last million years the Earth's climate has spent about 80 per cent of the time in glacial conditions. So the Amazon forest reconstructed for the last glacial period with a diverse mix of Andean and lowland tropic tree species and evergreen and partial evergreens is the norm. The lack of grassland in the Amazon during glacial periods also means we have to look for other evolution mechanisms for the huge diversity of the Amazon rainforest, and it may be that the ice ages were not the cause.

Unstable ice ages

In many ways glacial periods really should be called 'climate rollercoasters', because ice sheets are naturally unstable and during glacial periods the climate veers violently from one state to another as the ice sheets dramatically collapse and then reform. Most of the variations occur on a millennial time-scale, but the start of these extreme events can occur in as little as 3 years. The most impressive of these events are the Heinrich events. These events were named by Professor Wally Broecker, a palaeoceanographer at Lamont-Doherty Earth Observatory, after Hartmut Heinrich who wrote a paper describing them in 1988. Heinrich events are massive collapses of the North America Laurentide ice sheet that result in millions of tonnes of ice being poured into the North Atlantic Ocean. Wally Broecker described them as armadas of icebergs floating from North America across the Atlantic Ocean to Europe. Huge gouges have been found on the north French coast where these huge icebergs ran aground. These Heinrich events occurred against the general background of an unstable glacial climate, and represent the brief expression of the most extreme glacial conditions around the North Atlantic region. The Heinrich events are evident in the Greenland ice core records as a further 2–3 °C drop in temperature from the already cold glacial climate. The Heinrich events have been found to have had a global impact, with evidence of major climate changes described from as far afield as South America, the North Pacific, the Santa Barbara Basin, the Arabian Sea, the South China Sea, and the Sea of Japan. During these events around the North Atlantic region much colder conditions are found both in North America and Europe. In the North Atlantic Ocean the huge number of melting icebergs added so much cold fresh water that the sea-surface temperatures and salinity were reduced to the extent that surface water could not sink. This stopped all deep water formation in the North Atlantic Ocean switching off the global ocean conveyor belt.

Heinrich events are easy to spot in marine sediment cores from the middle of the Atlantic Ocean. This is because the icebergs bring huge amounts of rock with them into the ocean and as they melt they leave a trail of rock fragments scattered over the floor of the ocean. By recognizing these events in ocean sediments and dating the fossil in the sediment, it seems the Heinrich events occurred on average every 7,000 years during the last ice age. Also below these rock fragments we have found little burrows of marine worms. Usually these burrows cannot be seen as the sediment is mixed up by other animals coming to feed on it. For these fossil tubes and burrows to be preserved then a rain of rock fragments from the melting icebergs must have occurred within 3 years and been rapid enough to prevent other animals getting to the sediment. This evidence suggests that the collapse of the North American ice sheet was extremely rapid with icebergs flooding the Atlantic Ocean in less than 3 years. So, during an ice age conditions varied from cold conditions with massive ice sheets to extreme cold conditions brought on by the partial collapse of the North American ice sheet.

We now know that between the massive Heinrich events there are smaller events occurring about every 1,500 years, which are referred to as Dansgaard-Oescheger events or cycles. One suggestion is that Heinrich events are in fact just super Dansgaard-Oescheger events. The big difference between Heinrich events and Dansgaard-Oescheger events is that Heinrich events are only found during ice ages, while Dansgaard-Oescheger events have been found both in interglacial as well as in glacial periods.

What caused the Heinrich events?

The fascination with the Heinrich events is due to their occurrence on a time-scale that we can appreciate and the massive and profound effect they had on the climate of an ice age. So there are lots of competing theories for what caused them. A glaciologist called Doug MacAyeal suggested that the Heinrich event iceberg surges were caused by internal instabilities of the Laurentide ice sheet. This ice sheet rested on a bed of soft, loose sediment; when it is frozen it does not deform, behaving like concrete, and so would have been able to support the weight of the growing ice sheet. As the ice sheet expanded the geothermal heat from within the Earth's crust together with heat released from friction of ice moving over ice was trapped by the insulating effect of the overlying ice. This 'duvet' effect allowed the temperature of the sediment to increase until a critical point when it thawed. When this occurred the sediment became soft, and thus lubricated the base of the ice sheet causing a massive outflow of ice through the Hudson Strait into the North Atlantic. This, in turn, would lead to sudden loss of ice mass, which would reduce the insulating effect and lead to re-freezing of the basal ice and sediment bed, at which point the ice would revert to a slower build up and outward movement. Doug MacAyeal called this a binge–purge model and suggested all ice sheets have their own times of instability, thus the Fenno-Scandinavian, Greenland, and Icelandic ice sheets would have surges with different periodicities.

Another exciting theory is the 'bipolar climate see-saw' idea—another wonderful term created by Professor Wally Broecker. This theory is based on new evidence from the ice cores in Greenland and Antarctic, which show that during Heinrich events the climate of the Northern and Southern Hemispheres are out-of-phase. So when the climate of the Northern Hemisphere is cooling down it is warming up in Antarctica. It has been suggested that this so-called bipolar climate see-saw can be explained by alternating ice sheet collapse and resultant melt-water events in the North and South Atlantic Ocean. Each melt-water event would change the relative amount of deep-water formation in the two hemispheres and the resulting direction of the inter-hemispheric heat piracy. At the moment the Northern Hemisphere steals heat from the Southern Hemisphere to maintain the Gulf Stream and the relatively warm deep-water formation in the Nordic Seas. The heat is slowly returned by the flow of deep water from the North to the South Atlantic Ocean. So the bipolar climate see-saw model suggests if the ice sheets around the North Atlantic collapsed sending huge amounts of icebergs into the ocean they will melt. This melt water would make the ocean too fresh so none of the water could sink. This stops the formation of North Atlantic Deep Water and the Northern Hemisphere stops stealing heat from the Southern Hemisphere. This results in the Southern Hemisphere slowly warming up. Over maybe 1,000 years this heat build up is enough to collapse the edges of the Antarctic, which then shuts off the deep-water formation around Antarctica and the whole system is reversed. The nice thing about this theory is it can work in an interglacial period as well, and as we saw above the Dansgaard-Oescheger cycles of about 1,500 years occur during both glacial and interglacial periods.

Holocene

We are currently in an interglacial period called the Holocene, having escaped the last ice age about 10,000 years ago. The climate has not been constant during our interglacial period, and the early Holocene period may have been warmer and wetter than the 20th century. Throughout the Holocene there have been millennial-scale climate events, Dansgaard-Oescheger cycles, which involve a local cooling of 2°C. These events may have had a significant influence on classical civilizations. For example, the cold arid event about 4,200 years ago coincides with the collapse of many classical civilizations, including the Old Kingdom in Egypt; the Akkadian Empire in Mesopotamia; the Early Bronze Age societies of Anatolia, Greece, and Israel; the Indus valley civilization in India; the Hilmand civilization in Afghanistan; and the Hongshan culture of China. The last of these millennial climate cycles was the Little Ice Age. This event is really two cold periods; the first follows the Medieval Warm Period, which ended 1,000 years ago, and is often referred to as the Medieval Cold Period. The Medieval Cold Period played a role in extinguishing Norse colonies on Greenland and caused famine and mass migration in Europe. It started gradually before AD 1200 and ended at about AD 1650. The second cold period, more classically referred to as the Little Ice Age, may have been the most rapid and greatest change in the North Atlantic region during the late Holocene, as suggested by ice-core and deep-sea sediment records. In Britain the temperature dropped on average by 1°C—although everyone assumes it was much greater because of the beautiful paintings of the ice fairs on the frozen River Thames. But this is a myth because it would be nearly impossible now to get the weather in England cold enough to freeze the River Thames, as it is no longer a slow meandering river due to the demolition of the old London Bridge in 1831, the straightening of the River Thames to build the Embankment in the 1870s so Londoners could promenade like the Parisians, and the dredging of the river to make it an international port at the heart of the British Empire.

If we examine records around the globe it seems that the Little Ice Age and the Medieval Warm Period only occurred in northern Europe, northeast America, and Greenland. So the Little Ice Age was a regional climate fluctuation driven by small changes in the Gulf Stream and deep-water formation north of Iceland. Many climate change deniers suggest that global warming is just the world recovering from the Little Ice Age. But as most of the world never had a Little Ice Age there is nothing to recover from. The reconstructed global temperature records for the last 2 millennia are essential as they provide a context for the instrumental temperature data set for the last 150 years. It's clearly shown that temperatures have been warmer in the 20th and the 21st centuries than at any other time during the last two thousand years.

Summary

For the last 2.5 million years the climate of the Earth has been
dominated by the coming and going of the great ice sheets. These
ice sheets were so thick that there was two miles of ice piled up on
both North America and Northern Europe only 21,000 years ago.
The changes in global climate were profound. During a glacial
period the average global temperatures were 6°C lower than
today, global sea level was 120 m lower, and atmospheric carbon
dioxide was reduced by a third and atmospheric methane by a
half. The total weight of all the plants on the land was reduced by
as much as half. The landscape of the planet was dramatically
altered by the erosion and deposition of sediment by these huge
ice sheets. Major rivers were re-routed and mountains cut in half.
Land bridges appeared as the oceans lowered connecting
continents, allowing species to colonize new lands. It also seems
that over the last 2.5 million years the Earth's climate system
would much prefer to be in a cold state rather than the warm one
it is in today.

Chapter 8
Future climate change

Introduction

Future climate change is one of the defining challenges of the 21st century, along with poverty alleviation, environmental degradation, and global security. The problem is that 'climate change' is no longer just a scientific concern, it is now also of concern in terms of economics, sociology, geopolitics, national and local politics, law, and health, just to name a few. This chapter will examine briefly what 'anthropogenic' climate change is and the evidence that the global climate system is starting to change. The chapter will explain why changes in the climate system will lead to unpredictable weather patterns and an increase in the occurrence of extreme weather events, such as storms, floods, heatwaves, and droughts. Further details can be found in another of my titles in this series: *Global Warming: A Very Short Introduction*.

Human induced climate change

We have strong evidence that we have been changing the greenhouse gas content of the atmosphere. The first direct measurements of atmospheric carbon dioxide concentrations started in 1958 at an altitude of about 4,000 metres on the summit of Mauna Loa in Hawaii, a remote site free from local pollution. And looking even further back, air bubbles trapped in

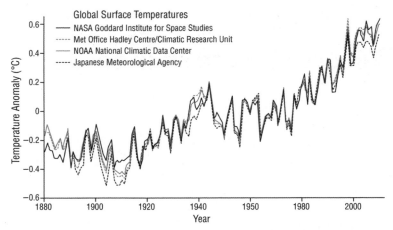

36. Global average surface temperature over the last 120 years

ice have been analysed from both the Greenland and the Antarctic ice sheets. These long ice core records suggest that pre-industrial carbon dioxide concentrations were about 280 parts per million by volume (ppmv). In 1958 the concentration was already 316 ppmv, and has climbed every year to reach 400 ppmv by June 2013. We have caused more pollution in one century than has occurred in thousands of centuries, throughout the natural waxing and waning of the great ice ages. Unfortunately, this increase in carbon dioxide in the atmosphere represents only half of the pollution we currently generate: about a quarter is absorbed by the oceans and another quarter by the land biosphere. One of the great worries scientists have is that this natural service may be reduced in the future making the situation worse.

According to the Intergovernmental Panel on Climate Change (IPCC) 2007 science report, increases in all greenhouse gases (see Chapter 2) over the last 150 years has already significantly changed the climate, including: an average rise in global temperatures of 0.75°C; a rise in sea level of over 22 cm; a significant shift in the seasonality and intensity of precipitation, changing weather patterns; and a significant retreat of Arctic sea ice and nearly all continental glaciers. According to National Aeronautics and Space Administration (NASA), National Oceanic and Atmospheric Administration (NOAA), the UK Meteorological Office, and the Japanese Meteorological Agency, in the last 150 years the last decade has been the warmest on record (see Figure 36). The IPCC in 2007 stated that the evidence for climate change is unequivocal and there is very high confidence that this is due to human activity. This view is supported by a vast array of expert organizations, including the Royal Society and American Association for the Advancement of Science.

The 'weight of evidence'

Understanding future climate change is about understanding how science works and the principle of the 'weight of evidence'. Science moves forward by using detailed observation and experimentation to constantly test ideas and theories. Over the last 30 years the theory of climate change must have been one of the most comprehensively tested ideas in science. First, as described earlier, we have tracked the rise in greenhouse gases in the atmosphere. Second, we know from laboratory and atmospheric measurements that these gases do indeed absorb heat. Third, we have tracked significant changes in global temperatures and sea level rises over the last century. Fourth, we have analysed physical changes in the Earth's system related to climate, including the retreating sea ice around the Arctic and Antarctica, retreating mountain glaciers on all continents, and the shrinking of the area covered by permafrost with an increase in depth of its active layer. The ice cover records from the Tornio River in Finland, compiled since 1693, show that the spring thaw of the frozen river now occurs a month earlier. Fifth, we have tracked weather records and seen significant shifts. In recent years massive storms and subsequent floods have hit China, Italy, England, Korea, Bangladesh, Venezuela, Pakistan, Australia, and Mozambique. These observations are supported by detailed compilations of all precipitation records for the Northern

Hemisphere published in *Nature* in 2011 by Dr Seung-Ki Min and his colleagues in Canada showing a significant increase in the intensity of rainfall over the last 60 years. Moreover, in Britain the winter of 2000/1 was the wettest six months since records began in the 18th century; August 2008 was the wettest August on record; and April–June 2012 was the wettest spring on record. Also data collected by the British public show that birds are nesting 12±4 days earlier than 35 years previously. Sixth we have analysed the effects of natural changes on climate including sun spots and volcanic eruptions and though these are essential in understanding the pattern of temperature changes over the last 150 years they cannot explain the overall warming trend. And, last, we now understand longer term past climate changes and the role greenhouse gases have played in setting the climate of our planet.

'Climategate'

Despite all the evidence, discussion of future climate change evokes strong reactions. In part because many of the changes we might have to make to ameliorate the situation seem to go against the current neo-liberal market driven approach in the West. It is also due to a fundamental misunderstanding of science by the media, the public, and our politicians. This is beautifully discussed in Mark Henderson's book *The Geek Manifesto*. 'Climategate' and the other supposed climate change cover-ups reported in the media are excellent examples of this misunderstanding. Because science is not a belief system, you cannot decide you believe in antibiotics (as they may save your life) and metal tubes with sticky out bits can safely fly you across the Atlantic Ocean, but yet deny smoking can cause cancer, or that HIV causes AIDS, or that greenhouse gases cause global warming. This is because science is a self-correcting, rational methodology based on collecting and building up evidence, which is at the very foundation of our society. In the case of 'Climategate' there was in November 2009 an illegal release, due to hacking, of thousands of emails and other documents from the University of East Anglia's (UEA) Climatic Research Unit (CRU). Allegations were made that the emails revealed misconduct within the climate science community including the withholding of scientific information, prevention of papers being published, the deletion of raw data, and the manipulation of data to make the case for global warming appear stronger than it is. Three independent inquiries concluded that there was no evidence of scientific malpractice. But what all the media commentators missed at the time was that two other major groups at NOAA and NASA had used different raw data sets and different statistical approaches and published the very same conclusions as the UEA group. This was further supported in 2012 when Professor Richard Muller, a physicist and previously a climate change sceptic, and his Berkeley group published their collated global temperature records for the last 200 years and he publicly announced that he had changed his mind and that climate change *was* occurring due to human activity.

Figure 37 shows all of these composite data sets for global temperature over the last 2000 years, and not surprisingly they are different but they show very similar trends and all suggest that the 20th century was warmer than any other time in the last two millennia.

There was also the accusation that the UEA group and by extension other climate scientist had changed the raw data. Short hand terminology used by scientists such as 'correct', 'trick', 'tweak', 'manipulate', 'a line', and 'correlate' of course did not help this very much. However, some raw data does need to be processed so it can be compared with other data, particularly if you are trying to make long records of temperature and the methods used to measure temperature have changed during that time. The clearest example of this is the measurement of sea temperature, which up to 1941 was made in sea water hoisted on deck in a bucket. Originally these buckets were wooden then

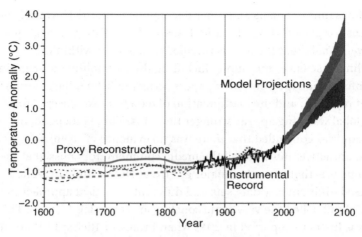

37. Past and future global average surface temperature

between 1856 and 1910 there was a shift to canvas buckets. This change in equipment would affect the level of cooling caused by evaporation as the water is being hoisted on deck. In addition, through this period there was a gradual shift from sailing ships to steamships, which altered the height of the ship decks and the speed of the ships, both of which could affect the evaporative cooling of the buckets. Since 1941, most sea temperature measurements have been made at the ships' engine water intakes—again, another shift. If scientists just stuck all this raw data together it would of course be wrong. Moreover in this case because the earlier sea-surface temperature measurements are too cold without correction it would make global warming in the ocean appear much greater than it really was. So the constant checking and correcting of data is extremely important in all parts of science. But the most important part is whether the results can be reproduced: is there a weight of evidence from many research groups showing the change? This is why after over 30 years of intensive research into climate change most scientists have a very high level of confidence that it is happening and it is due to human activity.

Climate change and its impact

In Chapter 3 we looked at how scientists model climate and future climate change. The synthesis of all the General Circulation Models is presented in the IPCC science reports. In 2007 they reported that global mean surface temperature could rise by between 1.1°C and 6.4°C by 2100, with best estimates being 1.8°C to 4°C. The biggest influence on how warm the climate becomes is that of emissions and what scenario is used. The faster and higher the greenhouse gas emissions rise, the hotter the world will become. It should be noted that global carbon dioxide emissions, despite the global recession, are rising as fast as the most dire 'business as usual' IPCC emission scenarios that have been looked at. The models also predict an increase in global mean sea level of between 18 cm and 59 cm. If the contribution from the melting of Greenland and Antarctica is included then this range increases to between 28 and 79 cm by 2100. All such predictions assume a continued linear response between global temperatures and ice sheet loss. This is unlikely, and sea level rise could thus be much higher. The next IPCC science report will be published in 2013 and will use more realistic future emission pathways, but draft chapters of this report show that they will reach very similar conclusions to the 2007 report.

The impacts of climate change will increase significantly as the temperature of the planet rises. The return period and severity of floods, droughts, heatwaves, and storms will increase. Coastal cities and towns will be especially vulnerable as sea levels rise, increasing the effects of floods and storm surges. A recent multi-disciplinary study by University College London published by the *Lancet* in 2009 demonstrated that the greatest threat of climate change to human health was from reduced water- and food-security, which could affect billions of people. Climate change also threatens the world's already devastated biodiversity. Ecosystems are already being hugely degraded by habitat loss, urbanization, pollution, and hunting. The 2007 'Millenium

Ecosystem Assessment' report suggested that three known species were becoming extinct each hour, while the 2008 Living Planet Index suggested that the global biodiversity of vertebrates had fallen by over a third in just 35 years, an extinction rate now 10,000 times faster than any observed in fossil records. The Royal Society's excellent 2012 'People and the Planet' report summarized the huge effects humans are having on the environment and how this will get worse as the global population increases but, more importantly, as consumption continues to rise uncontrollably around the world. Climate change of course will exacerbate all of this environmental degradation.

What is a 'safe' level of climate change?

So what level of climate change could be considered 'safe'? In February 2005 the British government convened an international science meeting in Exeter to discuss this very topic. This was so Tony Blair, the then prime minister, would have a political number to take to the G8 conference, which the UK was hosting later that year. Their recommendation was that global warming must be limited to a maximum of 2°C above pre-industrial average temperature. Below this threshold it seems that there were both winners and losers due to regional climate change, but above this figure everyone seems to lose. This, however, is a purely political perspective, because if you live on any of the low lying Pacific Islands then your whole island may have been flooded by the time we reach this 2°C level. However with the failure to produce a new climate treaty it now seems likely that temperature rises will exceed even this threshold. At the moment, in the 'business as usual' emissions scenario, we will hit that 2°C point long before 2050. This is not surprising given that the International Energy Authority has predicted that the use of fossil fuels over the next 20 years will include a 30 per cent increase in the use of oil, a 50 per cent increase in the use of coal, and a 40 per cent increase in the use of natural gas.

So what is the cost of saving the world? According to the UK government-commissioned Stern Review on 'The Economics of Climate Change' in 2006, if we do everything we can now to reduce global greenhouse gas emissions and ensure we adapt to the coming effects of climate change, it will only cost us 1 per cent of world GDP every year. However, if we do nothing then the impacts of climate change could cost between 5 and 20 per cent of world GDP every year. These figures have been disputed. Some experts have argued that the cost of converting the global economy to lower carbon emissions could cost more than 1 per cent GDP because global emissions have risen faster than the worst predictions. In response Sir Nicolas Stern has recently revised his figure to 2 per cent of world GDP. Others argue that the costs could be offset by regional carbon trading systems. Others suggest that the impacts and the associated costs of global warming have been under-estimated by IPCC and the Stern Review. However, even if the cost–benefit of solving global warming is less than suggested by the Stern Review, there is an clear ethical case for preventing the deaths of tens of millions of people and the increase in human misery for billions that is otherwise to come.

Despite the complete failure to produce a new global climate change treaty there are countries and regions taking the reports seriously. The UK has introduced the long-term legally binding Climate Change Act. This Act provides a legal framework for ensuring that government meets the target of reducing greenhouse gas emissions by at least 80 per cent by 2050, compared to 1990 levels. In spring 2012 the UK was joined by Mexico who have ratified their own climate change national law which will cut their emissions by 50 per cent by 2050. In the European Union all member countries have agreed to the '20:20:20' policy by the year 2020. This is a bid to achieve a 20 per cent cut in greenhouse gas emissions, a 20 per cent increase in energy efficiency, and for 20 per cent of all energy to be produced from renewable sources.

Summary

By 2030 global food and energy demand will have increased by 50 per cent and water requirements will have increased by 30 per cent. This is partly due to the rise in global population but mostly it is caused by the rapid development of lower income countries. Added to this is the growing effects of climate change, which directly threaten water and food security, and you have what Sir John Beddington (UK Government Chief Scientific Adviser) calls the 'perfect storm' (Figure 38). Hence climate change and sustainable energy are the key scientific issues of the 21st century. We already have clear evidence for climate change with a 0.75 °C rise in global temperatures and 22 cm rise in sea level during the 20th century. The IPCC predicts that global temperatures by 2100 could rise by between 1.8 °C and 4.0 °C, the range is due to the

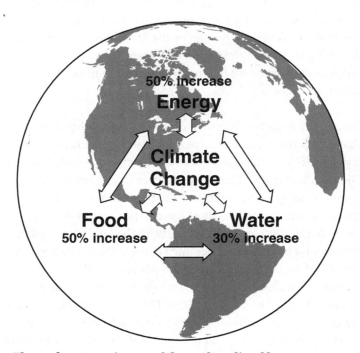

38. The perfect storm: increased demand predicted by 2030

uncertainty of how much greenhouse gas we will emit over the next 90 years. Sea levels could rise by between 28 cm and 79 cm—more, if the melting of the Greenland and Antarctica ice sheets accelerates. In addition, weather patterns will become less predictable and the occurrence of extreme climate events, such as storms, floods, heatwaves, and droughts, will grow more frequent. In the next chapter we will look at all the different options we have to fix our climate.

Chapter 9
Fixing climate change

Introduction

The most sensible approach to preventing the worst effects of future climate change would be to cut greenhouse gas emissions. Scientists believe a cut of between 50 per cent and 80 per cent in carbon dioxide by the middle of the century is required to avoid the worst effects. However, many have argued that the cost of significant cuts in fossil fuel use would severely affect the global economy. It would prevent rapid development and the alleviation of global poverty in a world where 8 million children die needlessly each year, 800 million people go to bed hungry each night, and 1,000 million people still do not have regular access to clean, safe drinking water. Nonetheless, under a business as usual emissions scenario there is a reasonable chance that we could be facing at least a 4°C warming by 2100, which would be disastrous and the impacts would fall disproportionally on the very poor in our society. In this chapter we will look at the three principal ways we could fix our climate. The first is 'mitigation' or reduction of the amount of carbon we emit. The second is the removal of carbon dioxide at source or from the atmosphere. The third is to use technology to reduce the amount of solar radiation being absorbed by the Earth, thus cooling the planet.

Mitigation

The idea of cutting global carbon emissions at least by half in the next 35 years and by up to 80 per cent by the end of the century may sound like fantasy, especially given the current trends in fossil fuel use (Figure 39). However, Steve Pacala and Robert Socolow, at Princeton University, published a very influential paper in the *Science* journal in a bid to make this challenge seem more achievable. They took the business as usual emissions scenario and the desired 450 parts per million by volume (ppmv) scenario and described the difference between the two as being represented by a number of 'wedges'. Instead of seeing one huge insurmountable problem, what we really have are lots of medium-sized changes, which would add up to the big change. They provided several examples for the wedges, each of them saving approximately 1 gigatonne of carbon every year. For example, one wedge would be doubling the efficiency of 2 billion cars from 30 miles per gallon (mpg) to 60 mpg, which actually is a very achievable aim, as there are family cars that have already been built that can easily do 100 mpg.

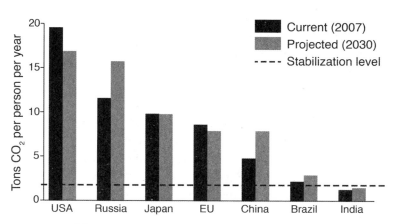

39. Past and future carbon emissions by country

One wedge is simply to improve energy efficiency. At the moment, the energy used in an average home in the USA is twice that of the average home in California; and the use in Californian domestic energy is twice that of the average home in Denmark. So already within the developed world there are huge savings to be made just by improving energy efficiency. If the industry and business sectors would reduce their energy use, they could significantly cut their running costs; however, this has not been an urgent priority for them due to the current incredibly low cost of energy. In any case, unfortunately it is likely that any efficiency gains these sectors made would ultimately be taken up by an increase in production and the level of energy use would increase accordingly. For example, if we did double the efficiency of 2 billion cars, this gain would be wiped out if production were increased by a further 2 billion cars. So, it is argued that the most important focus in cutting carbon dioxide production should be on the production of clean, or carbon-free, energy. This is discussed in the next section.

Alternative energy

The use of fossil fuels to provide energy has been an amazing discovery, allowing us to develop at a faster rate than at any time in history. The high standard of living in the developed world is based on the use of these cheap and relatively safe fossil fuels. Unfortunately, we have discovered that burning fossil fuels has had the unintended consequence of changing our global climate. This is because, as we burn these fossil fuels, we are releasing further solar energy that had been locked away millions of years ago. Instead, we need to switch from using energy generated in past climates to using energy from our current climate system. These include solar, wind, hydro-, wave, and tidal power. There are other reasons for us to switch to renewable sources of energy. First, there is the concern that we have reached the peak of our oil supplies and that the world is now running out of this fossil fuel. This may also be true for coal, despite there being hundreds of years' worth of high grade coal left. Second, countries have in the 21st century become very aware of 'energy security': most developed countries' economies are heavily reliant on imported fossil fuels, making them very vulnerable to volatility of the markets and international blackmail.

A brief discussion of the main alternative energy sources will now follow. It is immediately evident how certain sources of alternative energy would fit certain countries. For example, the UK has the best wind resource in the whole of Europe, while Saudi Arabia has excellent conditions for the sourcing of solar power. Almost all of the alternative sources use the climate system.

Solar energy

In Chapter 1 we discussed how the Earth receives on average 343 Watts per square metre (W/m^2) from the sun and that the Earth as a whole only received 2 billionths of all the energy put out by the sun. So the sun is in many ways the ultimate source of energy, which plants have been utilizing for billions of year. At the moment we can convert solar energy directly for heating or electricity, or we can capture the energy through photosynthesis by growing biofuels. The simplest approach is that of 'solar heating'. On a small scale, houses and other buildings in sunny countries could have solar heating panels on the roofs, which heat up water, for example, so people can have carbon-free hot showers and baths. On a large scale, parabolic mirrors can be used to focus solar energy to generate hot liquid (water or oil) with which to drive turbines to create electricity. The best places to situate solar heat plants are in low-latitude deserts, which have very few cloudy days per year. Solar heat plants have been built and used in California since the 1980s and are now being built and used in many other countries. Solar photovoltaic or solar panels convert sunlight directly into electricity. The individual rays of the sun hit the solar panel and dislodge electrons inside it, creating an electrical current. The main advantage of solar panels is that you can place them where the energy is needed and avoid the complicated infrastructure normally required to move electricity around. Over the last decade there has been a massive increase in their efficiency, the best commercially available solar panels are about 21 per cent efficient, which is great compared with photosynthesis that is about 1 per cent. There has been a significant drop in price in these solar panels with huge investment in the technology.

Biofuels

Biofuels are generated by the conversion of solar energy into plant biomass via photosynthesis, which can then be used to produce a liquid fuel. The global economy is based on the use of liquid fossil fuels, particularly for the transport sector. So in the short-term fuels derived from plants could be an intermediate low carbon way of powering cars, ships, and aeroplanes. There is the problem that the production of biofuels could compete with that of food crops. Indeed, the food price peaks of 2008 and 2011 were initially blamed on biofuel production. However, analysis by the New England Institute of Complexity showed these massive increases in price >50 per cent were in fact due to speculation on food prices in the financial markets.

Ultimately, electric cars are the future, because we can guarantee that the electricity produced is carbon-neutral. However, this is not an option for aeroplanes. Traditional airplane fuel is kerosene, which combines relative light weight with a high energy output. Research is being carried out to see whether a biofuel could be produced that would be light enough and powerful enough to replace kerosene.

Wind energy

Wind turbines are an efficient means of generating electricity—if they are large enough and preferably located out at sea. Ideally, you want turbines the size of the Statue of Liberty for maximum effectiveness. For example, the London Array is being built in the River Thames estuary and will generate 1,000 Megawatts (MW) of electricity, making it the world's largest consented offshore wind farm. When finished it could power up to 750,000 homes—about a quarter of Greater London—and reduce harmful carbon dioxide emissions by 1.4 million tonnes a year.

There are problems with wind turbines. First, they do not represent a constant supply of electricity: if the wind does not blow, then there is no electricity. Second, people do not like them, because they think they are ugly, noisy, and worry about the effects on local natural habitats. All these problems are easy to overcome by situating wind farms in remote locations, such as out at sea and away from areas of special scientific or natural interest. One study suggests that wind in principle could generate over 125,000 Terrawatt-hours, which is five times the current global electricity requirement.

Wave and tidal energy

Wave and tidal power could be an important source of energy in the future. The concept is simple, to convert the continuous movement of the ocean in the form of waves into electricity. However, this is easier said than done, and experts in the field suggest that wave power technology is now where solar panel technology was about 20 years ago—a lot of catching up is required. But tidal power in particular has one key advantage over solar and wind power: it is constant. In any country for the energy supply to be maintained there has to be at least 20 per cent production guaranteed, this is referred to as 'the baseline'. With the switch to alternative energy, new sources of power to ensure this constant baseline level need to be developed.

Hydroenergy

Hydroelectric power is an important source of energy globally. In 2010, it supplied 5 per cent of the world's energy. The majority of the electricity comes from large dam projects. These projects can present major ethical problems as large areas of land must be flooded above the dam, causing mass relocation of people and the destruction of the local environment. A dam also slows the flow of water in a river affecting the depositing of nutrient-rich silt in areas lower down on the river's course. If the river crosses national boundaries, there are potential issues over the rights to water and silt. For example, one of the reasons why Bangladesh is sinking is due to the lack of silt, and this lack has been caused by dams on the major rivers in India. There is also a debate about how much greenhouse gas hydroelectric plants save, because, even though the production of electricity does not cause any carbon emissions, the rotting vegetation in the area that has been flooded behind the dam does give off significant amounts of methane.

There are other sources of alternative or low-carbon energy which do not rely on the climate and these are briefly discussed in the next section, for completeness.

Geothermal energy

Below our feet, deep within the Earth, is hot, molten rock. In some locations, for example in Iceland and Kenya, this hot rock comes very close to the Earth's surface and can be used to heat water to make steam. This is an excellent, carbon-free source of energy, because the electricity generated from the steam can be used to pump the water down to the hot rocks. Unfortunately, this energy source is limited by geography. There is, however, another way this warmth of the Earth can be used. All new buildings could have a borehole below them with ground-sourced heat pumps. In this system, cold water is pumped down into these boreholes and the hot, molten rock below heats the water, cutting the cost of heating water, and this technology could be used almost everywhere in the world.

Nuclear fission

Energy is generated when you split heavy atoms such as uranium; this is known as nuclear fission. It has a very low direct carbon signature, however a significant amount of carbon is generated both from mining the uranium and from later decommissioning the power station. At the moment, 5 per cent of global energy comes from nuclear power. The new generation of nuclear power stations is extremely efficient: a level of nearly 90 per cent efficiency. The main disadvantages of nuclear power relate to the generation of high-level radioactive waste and the issue of safety. However, improvements in efficiency have reduced waste and the new generations of nuclear reactor sites have state-of-the-art safety precautions built in. The advantages of nuclear power are that it is reliable and can provide the required baseline in the energy mix; furthermore, it is a technology that is ready to go and has already been thoroughly tested.

Nuclear fusion

Nuclear fusion is the generation of energy when two smaller atoms are fused together. This is what happens in our sun and every other star. The idea is that the heavy form of hydrogen found in sea water can be combined and the only waste product is the non-radioactive gas helium. The problem, of course, is persuading those two atoms to join together. The sun does it by subjecting the atoms to incredibly high temperatures and pressures. Some advances have been made at the Joint European Torus (JET) project in the UK, which has produced 16 MW of fusion power. The problem is the amount of energy required to generate the huge temperatures in the first place and the difficulty of scaling the production up to power plant level.

Carbon dioxide removal

'Geoengineering' is the general term used for technologies that could be used to either remove greenhouses gases from the atmosphere or to change the climate of the Earth (Figure 40). Ideas considered under geoengineering range from the very sensible to the completely mad. At the moment we currently release over 8.5 gigatonnes of carbon (GtC) per year, so any interventions must be able to operate at a very large scale. In this section we will consider removal and disposal of atmospheric

40. Geoengineering summary

carbon dioxide. There are three main approaches: biological, physical, and chemical.

Biological carbon removal

Biological approaches to carbon removal include the use of biofuels, discussed earlier, and reforestation. Deforestation or afforestation and avoidance of deforestation are sensible win-win solutions. By maintaining our forests we can retain biodiversity, stabilize soils and local rainfall, and provide livelihoods for local people via carbon credits. An excellent example of this can be seen in China. By 1990 the Loess Plateau, the bread-basket for China for at least the last 3,000 years, was turning into a dust bowl. Reforestation and overworking of the soils had started to reduce fertility, so farmers were cutting down more trees to open up more land to produce enough food to survive. The Chinese government became aware of this problem and over-the subsequent 10 years instigated a radical tree-planting programme, with severe punishments for anyone caught chopping down trees. The effects were amazing, the trees stabilized the soils, greatly reducing soil erosion. The trees added moisture to the atmosphere, through transpiration, reducing evaporation and water loss. Once the forest reached a critical size and area it also started to stabilize the rainfall. In Chapter 8 we saw that the land's biosphere is already absorbing about 2 GtC per year of our pollution and Steve Pacala and Robert Socolow estimate that if we completely stopped global deforestation and doubled our current rate of planting we could produce another one of their 1 GtC per year wedges with all the win-win benefits that go with reforestation. In the UK the Forestry Commission proposed to increase the forested land from 12 per cent to 16 per cent by 2050. This would mean the government's target of 80 per cent reduction in carbon dioxide emissions by 2050 would only be 70 per cent, thanks to the absorption and storage of carbon by our forests.

The second biological approach is changing the uptake of carbon dioxide by the oceans. The most famous ocean 'technofix' was suggested by the late Professor John Martin. He suggested that many of the world's oceans are under-producing. This is because of the lack of vital micro-nutrients, the most important of which is iron, which allows plants to grow in the surface waters. Marine plants need minute quantities of iron and without it they cannot grow. In most oceans enough iron-rich dust gets blown in from the land, but it seems that large areas of the Pacific and Southern Ocean do not receive much dust and thus are barren of iron. So it has been suggested that we could fertilize the ocean with iron to stimulate marine productivity. The extra photosynthesis would convert more surface-water carbon dioxide into organic matter. Furthermore, when the organisms die, the organic matter drops to the bottom of the ocean, taking with it and storing the extra carbon. The reduced surface-water carbon dioxide is then replenished by carbon dioxide from the atmosphere. So, in short, fertilizing the world's oceans could help to remove atmospheric carbon dioxide and store it in deep-sea sediments. Experiments at sea have produced highly variable results, with some showing no effects at all while others revealing that the amount of iron required would be huge. Also, as soon as you stop adding the extra iron, most of this stored carbon dioxide is released, as very little organic matter is allowed to escape out of the photic zone (the part of the ocean close enough to the surface to receive the necessary sunlight for photosynthesis to take place) per year.

Physical

Removal of carbon dioxide during industrial processes is tricky and costly, because not only does the carbon dioxide need to be removed, but it must be stored safely. Removal and storage costs could be somewhere between $10 and $50 per tonne of carbon dioxide. This would cause a 15 per cent to 100 per cent increase in power production costs. However, due to the cheap and plentiful supply of high-grade coal, carbon capture and storage (CCS) is held to be one of the greatest hopes of governments worldwide. A lot more research is needed in this area to make CCS work efficiently and also to get it to an affordable level. The problem though is non-CCS coal or gas produced electricity will always be cheaper than CCS, so legislation is required to ensure companies are obliged to use CCS or to switch to alternative energy production. For example, the European Union (EU) emissions trading scheme (ETS), a 'cap-and-trade' system that all companies producing or using large amounts of energy are required to be part of, is helping reduce the EU total emissions by 2020.

Another possible solution is removal of carbon dioxide directly from the air. Considering carbon dioxide makes up just 0.04 per cent of the atmosphere this is much harder than it sounds. One mad idea is the production of artificial or plastic trees. Klaus Lackner a theoretical physicist and Allen Wright an engineer, supported by Wally Broecker, a climatologist, have designed carbon dioxide-binding plastic which can scrub carbon dioxide out of the atmosphere. In the proposed system, the carbon dioxide is then released from the plastic and taken away for storage. The first problem is water, as the plastic releases the carbon dioxide into solution when wet. This means that the plastic trees would have to be placed in very arid areas or would require giant umbrellas to protect them from rainfall. The second problem is the amount of energy that would be required to build, operate, and then store the carbon dioxide. The third problem is one of scale: tens of millions of these giant artificial trees would be required just to deal with US carbon emissions. When debating this approach on radio, I very gently suggested that perhaps we could just plant normal trees.

However, if plastic trees are not the answer, another form of technology to remove carbon dioxide either at source or ultimately from the atmosphere may be required.

Weathering

Carbon dioxide is naturally removed from the atmosphere over hundreds and thousands of years through the process of weathering. This process was discussed in Chapter 6 when we discussed the role of the uplift of the Himalayas in removing carbon dioxide from the atmosphere. Atmospheric carbon dioxide (CO_2) can interact with calcium silicate $(CaSiO_3)$ directly:

$$CaSiO_3 + CO_2 \rightarrow CaCO_3 + SiO_2$$

This is an extremely slow process removing less than 0.1 GtC per year, which is a hundred times less than we are emitting. Another process uses the combination of rainwater and carbon dioxide to form a weak carbonic acid solution:

$$CaSiO_3 + 2CO_2 + H2O \rightarrow Ca^{2+} + 2HCO_3^- + SiO_2$$

Only weathering of silicate minerals makes a difference to atmospheric carbon dioxide levels, as weathering of carbonate rocks by carbonic acid returns carbon dioxide to the atmosphere. By-products of hydrolysis reactions affecting silicate minerals are biocarbonates (HCO_3^-), metabolized by marine plankton and converted to calcium carbonate. The calcite skeletal remains of the marine biota are ultimately deposited as deep-sea sediments and hence lost from the global biogeochemical carbon cycle for the duration of the lifecycle of the oceanic crust on which they were deposited.

There are a number of geoengineering ideas aimed at enhancing these natural weathering reactions. One suggestion is to add silicate minerals to soils that are used for agriculture. This would remove atmospheric carbon dioxide and fix it as carbonate minerals and biocarbonate in solution. However the scale at which this would have to be done is very large and the effects on soils and their fertility are unknown. Another suggestion is to enhance the rate of reaction of carbon dioxide with basalts and olivine rocks in the Earth's crust. Concentrated carbon dioxide would be injected into the ground and would create carbonates deep underground. Again, like many geoengineering ideas, this is a

Climate

great suggestion but very little work has been done to see if it is feasible, safe, and possible to do on the necessary scale.

Storage

Not all recovered carbon dioxide has to be stored; some may be utilized in enhanced oil recovery, the food industry, chemical manufacturing (producing soda ash, urea, and methanol), and the metal-processing industries. Carbon dioxide can also be applied to the production of construction material, solvents, cleaning compounds, packaging, and in waste-water treatment. But in reality, most of the carbon dioxide captured from industrial processes would have to be stored. It has been estimated that theoretically two-thirds of the carbon dioxide formed from the combustion of the world's total oil and gas reserves could be stored in the corresponding reservoirs. Other estimates indicate storage of 90–400 gigatonnes in natural gas fields alone and another 90 gigatonnes in aquifers.

Oceans could also be used to dispose of the carbon dioxide. Suggestions have included storage by hydrate dumping—if you mix carbon dioxide and water at high pressure and low temperatures, it creates a solid, or hydrate, which is heavier than the surrounding water and thus drops to the bottom. Another more recent suggestion is to inject the carbon dioxide half a mile deep into shattered volcanic rocks in between giant lava flows. The carbon dioxide will react with the water percolating through the rocks. The acidified water will dissolve metals in the rocks, mainly calcium and aluminium. Once it forms calcium bicarbonate with the calcium, it can no longer bubble out and escape. Though if it were to escape into the ocean, then bicarbonate would be relatively harmless. With ocean storage there is the added complication that we saw in Chapter 2: the ocean circulates, so whatever carbon dioxide you dump, some of it will eventually return. Moreover, scientists are very uncertain about the environmental effects this would have on the ocean ecosystems.

The major issue with all of these methods of storage is that of safety. Carbon dioxide is a very dangerous gas because it is heavier than air and can cause suffocation. An important example of this was a tremendous explosion of carbon dioxide from Lake Nyos that occurred in 1986, in the west of Cameroon, killing more than 1,700 people and livestock up to 25 km away. Though similar disasters had previously occurred, never had so many people and animals been asphyxiated on such a scale in a single brief event. What scientists now believe happened was that dissolved carbon dioxide from the nearby volcano seeped from springs beneath the lake and was trapped in deep water by the weight of the water above. In 1986, there was an avalanche that mixed up the lake waters, resulting in an explosive overturn of the whole lake, and all the trapped carbon dioxide was released in one go. However, huge amounts of mined ancient carbon dioxide are constantly being pumped around the USA to enhance oil recovery. There are no reports of any major incidents and engineers working on these pipelines feel they are much safer than they would be on gas or oil pipelines, which run across most major cities.

Solar radiation management

As we have seen, many of the geoengineering solutions proposed are still just ideas and need a lot more work to see if they are feasible. This is even more true of the solar radiation management ideas that have been proposed, many of which sound like something from a bad Hollywood B-movie. These suggestions include changing the albedo of the Earth: increasing the amount of solar energy reflected back into space to balance the heating from global warming (Figure 40). Methods to achieve this include: erecting massive mirrors in space; injecting aerosols into the atmosphere; making crops more reflective; painting all roofs white; increasing white cloud cover; and covering large areas of the world's deserts with reflective polyethylene-aluminium sheets. The fundamental problem with all of these approaches is that we have no idea what knock-on effects they would have. At the moment, we are performing one of the largest geoengineering experiments ever undertaken, by injecting huge amounts of greenhouse gases into the atmosphere, and though we have some idea of what might happen, we have no idea what the specific effects on our climate system could be. This is equally true of these geoengineering solutions—we currently have little idea if they would work or what unaccounted-for side-effects they might have. In many ways, climate change for the Earth can be seen in the same way as illness and the human body: it is always preferable to prevent an illness than to try and cure one, and we all know the potential side effects of drugs and chemo- or radiation therapy.

Let us use just one of these more far-fetched ideas as an example of what is wrong with solar radiation management: mirrors in space to deflect the sunlight. The most sophisticated of these suggestions is from Roger Angel, Director of the Centre for Astronomical Adaptive Optics at the University of Arizona, who suggests the use of a mesh of tiny reflectors to bend some of the light away from the Earth. He himself admits this would be expensive, requiring 16 trillion gossamer-light spacecraft costing at least $1 trillion and taking 30 years to launch. But, like all the other ideas we have discussed intended to change the Earth's albedo, it will not work. The reason for this is that all these approaches are fixated with getting the Earth's average temperature down, which misses out the importance of current temperature distribution, which as we know from Chapter 2 is what drives our climate. In fact, using climate models, Dan Lundt and colleagues at Bristol University show that these approaches would take us to a completely different global climate: the tropics would be 1.5°C colder; the high latitudes would be 1.5°C warmer; and precipitation would drop by 5 per cent globally compared with pre-industrial times.

Geoengineering governance

The Royal Society 2009 report on geoengineering not only reviewed the current scientific material in this area but also made the important step of trying to understand the governance issues associated with playing with the global climate system. There are a large number of ethical issues when considering how changing regional and global climate may affect countries differently. There may be overall positive results but minor changes in rainfall patterns could mean whole countries not receiving enough or too much rain, possibly resulting in disasters. The Royal Society summarized three main views on geoengineering:

1. It is a route to buy back some time to allow the failed international mitigation negotiations to catch up.
2. It represents a dangerous manipulation of the Earth's system and may be intrinsically unethical.
3. It is strictly an insurance policy against major mitigation policy failure.

Even if research is allowed to go ahead and geoengineering solutions are required, as with many emerging areas of modern technology, new flexible governance and regulatory frameworks will be required. Currently there are many international treaties which have a bearing on geoengineering and it seems that no single instrument applies. Hence 'fixing' our climate challenges our nation-state view of the world and new ways of governing will be required in the future.

Adaptation

Even if we decided to reduce carbon emissions significantly and tried all the geoengineering options available there would still be some climate change. This is because there has already been a temperature increase of 0.76°C, and even if we could reduce atmospheric carbon dioxide concentration to 2000 levels that would still add at least another 0.6°C. Considering the failure of international talks and the lack of serious investment in geoengineering we are currently on a 'business as usual' pathway. This means that many countries will be adversely affected by climate change in the near future, and nearly all countries will be affected in the next 30 years. So our failure to fix our climate issues means we must also plan to adapt to a new, changing climate. Each national government needs to research the vulnerabilities of their environment and socioeconomic system and predict the most likely effects of climate change for them.

The major threat from climate change, however, is its unpredictability. Humans can live in almost any extreme of climate from deserts to the Arctic, but only when we can predict the range of that extreme weather. So adaptation is really about how each country or region can deal with emerging threats of new levels of extreme weather. This adaptation should start now, as infrastructure changes, particularly in democracies, can take over 30 years to implement. For example, if you want to change land-use by building better sea defences or by changing farmland back to natural wetlands in a particular area, it can take up to 20 years to research and plan the appropriate measures. It can then take another 10 years for the full consultative and legal processes to be completed; a further 10 years to implement these changes; and another 10 years for the natural restoration to take place. A good example of this is the Thames Barrier, which currently protects London from flooding. It was built in response to the severe flooding in 1953, but it did not open officially until 1984—31 years later.

The other problem is that adaptation requires money to be invested now; and many countries just do not have the money. People do not want to pay more taxes to protect themselves in the future since most people live for today. Despite the fact that all of the adaptations discussed will in the long-term save money for the local area, the country, and the world, we as a global society still have a very short-term view—usually measured in the few years that elapse between successive governments.

The one thing that every government could do immediately is to set up a climate change impact assessment. For example, in the

UK there is the UK Climate Impact Programme (<http://www.ukcip.org.uk>), which shows the possible effects of climate change on the UK over the next 100 years. This programme is aimed at the UK national and local government, industry, business, the media, and the general public. If every government set up one of these programmes then at least their citizens would have the information to make informed choices about how their countries should be adapting to climate change.

Summary

So how should we fix global climate change? First it seems sensible to have an international political solution. We are currently without a post-2012 agreement and are looking at huge increases in global carbon emissions (Figure 41). Any political agreement will have to include measures to protect the rapid development of developing countries. It is a moral imperative that people in the poorest countries have the right to develop and to obtain the same lifestyle enjoyed by the developed world. We also need massive investment in alternative/renewable power sources and low carbon technology to provide the means of reducing world carbon emissions. There should be investment in geoengineering solutions, especially those that will make a significant impact in the short-term, such as reforestation and CCS. Action on climate change should also always contain an element of win-win. For example, supporting a huge increase in renewable energy not only reduces emissions but helps to provide energy security by reducing the reliance on imported oil, coal, and gas. Reduced deforestation and reforestation should not only draw down carbon dioxide from the atmosphere but will also help to retain biodiversity, stabilize soils, and provide livelihoods for local people via carbon credits. Measures that reduce car use will increase walking and cycling, which in turn should improve people's health, for example reducing levels of obesity and heart attacks.

We must not pin all our hopes on global politics, clean energy technology, and geoengineering—we must prepare for the worst

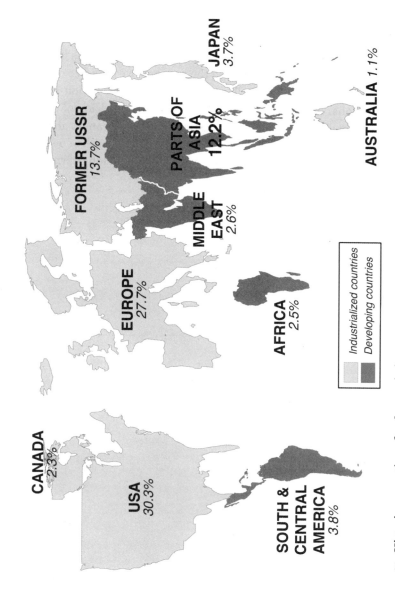

CANADA
2.3%

USA
30.3%

SOUTH &
CENTRAL
AMERICA
3.8%

EUROPE
27.7%

FORMER USSR
13.7%

MIDDLE
EAST
2.6%

PARTS OF
ASIA
12.2%

AFRICA
2.5%

JAPAN
3.7%

AUSTRALIA 1.1%

Industrialized countries
Developing countries

41. Historic proportion of carbon emissions

and adapt. If implemented now, much of the cost and damage that could be caused by changing climate can be mitigated. However, this requires nations and regions to plan for the next 50 years, something that most societies are unable to do due to the very short-term perspective of political institutions. This means that our climate issues are challenging the very way we organize our society. Not only do they challenge the concept of the nation-state versus global responsibility, but also the short-term vision of our political leaders. To answer the question of what we can do about climate change—we must change some of the basic rules of our society to allow us to adopt a much more global and long-term sustainable approach.

Chapter 10
Ultimate climate change

Introduction

By understanding how the climate of the Earth has varied in the past it is possible to look into the future. Many of the processes of change introduced in this book can be played forward to see what will happen in the near and far future.

The next ice age

The latest scientific research tells us the slow descent into a new ice age should start anytime in the next 1,000 years. As we saw in Chapter 7 we understand how the different wobbles of the Earth's orbit affect the Earth's climate. We can also look at previous interglacial periods to see how long they have lasted. Chronis Tzedakis, a professor of palaeoclimatology at University College London, and colleagues have calculated the natural length of the Holocene period and shown that any time from now until 1,500 years in the future the next ice age should start. But they also concluded that it probably will not happen. It seems that during every other interglacial period atmospheric carbon dioxide is highest as the climate system rebounds out of an ice age. Atmospheric carbon dioxide slowly lowers through each interglacial period until it hits a critical value of about 240 parts per million by volume (ppmv), which is 40 ppmv lower than pre-industrial times and 160 ppmv lower than today. Once this critical value has been reached then the climate system can respond to orbital forcing and slide into the next glacial period. However, it seems that our atmospheric carbon dioxide pollution, if it remains high, will prevent the world from sliding into an ice age over the next 1,500 years. In fact, according to the model predictions by Professor Andre Berger at the Université Catholique de Louvain in Belgium, if we double atmospheric carbon dioxide concentrations then global warming would postpone the next ice age for another 45,000 years. Interestingly, however, by that time the orbital forcing will be great enough to overcome extreme levels of greenhouse gases and normal glacial–interglacial cycles will reassert themselves.

Another interesting point is, why, even before the Industrial Revolution, was the atmospheric carbon dioxide level already above where we would have predicted it should be? This brings in the wonderful Ruddiman early anthropocene hypothesis. Bill Ruddiman, a professor of palaeoclimatology at the University of Virginia, has proposed that early agriculturalists caused a reversal in natural declines of atmospheric carbon dioxide starting about 7,000 years ago and of atmospheric methane starting about 5,000 years ago. This argument has caused huge controversy, but like all good theories it has been tested again and again, and no one has yet been able to disprove it. So essentially the argument is that early human interactions with the environment increased atmospheric greenhouse gases just enough that even prior to the Industrial Revolution there was enough of a change to the climate to delay the onset of the next ice age. It also raises the question of when exactly humans became such a geological force; and since all the ages of geology are given a name and there is a movement to define the age in which humans began to have a significant effect on the Earth's climate system.

The late ecologist Eugene F. Stoermer coined the term the 'Anthropocene', which has been popularized by the Nobel

Prize-winning atmospheric chemist Paul Crutzen. The term has Greek roots: *anthropo-* meaning 'human' and *-cene* meaning 'new'. Though the term is gaining a lot of support we still do not know where to put the boundary between the Holocene and the Anthropocene. In geology, boundaries between periods have to have a well defined datum or 'golden spike' in order to be recognized worldwide. Some have argued for Bill Ruddiman's proposed period of early human influence on atmospheric carbon dioxide, others have argued for the Industrial Revolution, and still others have argued for the period to be dated using trace element evidence of civilization (for example, the layer of chlorine from 1960s atomic weapon testing programmes) which can be found in ice cores. Whatever datum is finally chosen, there is no doubt that humans are now a major 'geological' force on the Earth: changing global patterns of erosion through massive land-use changes, including deforestation, which have caused mass extinction and huge loss of biodiversity, altering the global nitrogen cycle, ozone depletion, and of course climate change.

The next supercontinent

Christopher Scotese a professor at the University of Texas at Arlington is the director of the PALEOMAP Project. The project aims to illustrate how plate tectonics has changed ocean basins and continents and their positions over the last billion years. They also speculate how plate tectonics will change the face of the Earth in the future. In Chapter 5 we saw how supercontinents formed in the past and the severe effect they had on both climate and evolution. According to the PALEOMAP Project, the next supercontinent will form in the next 250 million years. Up to 50 million years in the future the world looks similar but with a few changes: the Atlantic Ocean will continue to widen; Africa will collide with Europe closing the Mediterranean Sea; Australia will collide with Southeast Asia; and California will slide northwards up the coast to Alaska (Figure 42). The really interesting changes happen between 50 and 150 million years in the future. Key to these are the major changes along the east coast of North and South America. Currently this is a passive margin and the continent and ocean plates are joined. But sometime around 50 million years in the future the Atlantic continents will pass over the Caribbean and Cocos plate boundary forming a new subduction zone (where tectonic plates collide, pushing one below the other) in the

Climate

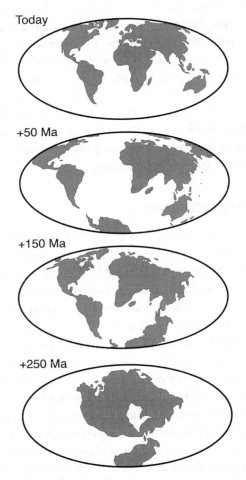

Today

+50 Ma

+150 Ma

+250 Ma

42. Future locations of the continents

Atlantic. This will create the Western Atlantic trench where the Atlantic Ocean continental plates will start to be consumed. Despite the mid-Atlantic ridge continuing to build new oceanic plate material, eventually the destruction will overtake the creation and the Atlantic Ocean will start to close up. At about 150 million years into the future the mid-Atlantic ridge will reach the subduction zone and be consumed. With no new ocean crust being created the closure of the Atlantic Ocean will then speed up. Elsewhere, the UK and Europe will then have a North Pacific Ocean view; the Mediterranean mountains will have reached their maximum height; and Antarctica and Australia will form one large continent. By 250 million years in the future the Americas, Africa, and Asia will have joined to form one supercontinent, with an inland sea the size of Australia. A small ocean gateway will separate this supercontinent from the Antarctic-Australian continent. We know from palaeoclimate records that supercontinents are bad for life. For example 96 per cent of all marine species and 70 per cent of terrestrial vertebrate species became extinct during the supercontinental Permian–Triassic extinction event 250 million years ago. So in another 250 million years, whatever life exists on Earth will face another major challenging period.

Boiling oceans

We assume the cycles of supercontinents forming and fragmenting will continue for as long as there are continents. However, there is a point where the climate will deteriorate to the point where complex, multi-cellular organisms cannot survive. For example, since the formation of our sun it has been increasing its energy output. The sun's luminosity increases about 10 per cent every billion years. Professor James Lovelock has suggested that feedbacks between life and climate system have modified the greenhouse gas content of the atmosphere to take account of this increase over the last few billion years; which is a central argument of his wonderful Gaia hypothesis. However in geological terms we already have extremely low levels of greenhouse gases, despite our best efforts over the last 100 years to reverse this. So the planet's ability to cool itself down is approaching its limit. This means that it is likely that over the next few billion years the temperature of the Earth will creep up (Figure 43). There will come a critical point when it is so warm that the oceans will start to evaporate, pumping huge amounts of moisture into the atmosphere. As we saw in Chapter 2 water vapour is one of the most important greenhouse gases and this runaway greenhouse would take average temperatures on Earth to over 100 °C, which no current multi-cellular organism can survive. I would suggest that ultimate climate change will occur when the oceans boil causing super global warming. It is rather ironic that the very place where life, and then complex life, evolved will be its ultimate destructor. Even the most extreme microbes will only last another 3 billion years and then it will be too hot even for them (Figure 43).

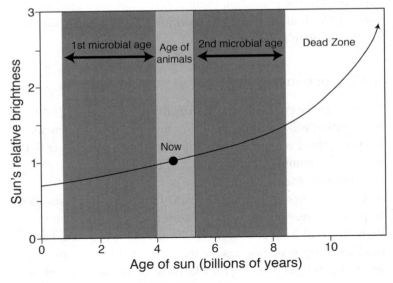

43. Life on Earth over the life of the sun

Death of the Earth

While multi-cellular life on Earth is likely to end in about a billion years' time and microbes in about 3 billion years' time, this would not be the end of the Earth. That is predicted to happen in about 5 billion years time. The sun is about halfway through turning its hydrogen into helium. Each second, more than 4 million tonnes of matter are converted into energy within the sun's core. Over the last 4.5 billion years the sun has turned 100 Earth-masses of matter into energy. The sun, however, does not have enough mass to explode as a supernova. Instead, in about 5 billion years it should enter a red giant phase. Its outer layers should expand as the hydrogen fuel at the core is consumed and the core should contract and heat up. Hydrogen fusion should then continue along a shell surrounding a helium core, which should steadily expand as more helium is produced. Once the core temperature reaches around 100 million °C the sun should even start producing carbon through the consumption of helium. As a red giant, the sun would be bigger than the Earth's current orbit area, about 250 times the present radius of the sun. It was thought that the Earth might survive even this, as the sun would have lost roughly 30 per cent of its present mass by then, so its gravity would be less strong, resulting in the orbits of the surrounding planets moving outwards. However, Peter Schröder of the University of Guanajuato in Mexico and Robert Smith of the University of Sussex in the UK created a detailed model of the sun's transition to a red giant and they found that the Earth's orbit would widen at first, but then the Earth would itself cause a 'tidal bulge' on the sun's surface from its own gravitational pull. The bulge would lag just behind the Earth in its orbit, slowing it down enough to drag it to a fiery demise. After the red giant phase and the death of the Earth, the sun would throw off its outer layers, forming a planetary nebula. The only object that would remain after the outer layers were ejected would be the extremely hot stellar core, which would slowly cool and fade over many billions of years to become a white dwarf star.

Further reading

Science of climate and weather

R. G. Barry and R. J. Chorley, *Atmosphere, Weather and Climate*, Routledge, 9th edition, p. 536 (2009)

A. Colling (ed.), *The Earth and Life: The Dynamic Earth*, Open University Worldwide, p. 256 (1997)

J. Gleick, *Chaos: Making a New Science*, Vintage, new edition, p. 380 (1997)

R. Hamblyn, *The Cloud Book: How to Understand the Skies*, David & Charles Publishers, p. 144 (2008)

T. R. Oke, *Boundary Layer Climates*, Routledge, second (reprinted) edition, p. 464 (2001)

Past climate

R. B. Alley, *The Two-Mile Time Machine: Ice Cores, Abrupt Climate Change, and Our Future*, Princeton University Press, new edition, p. 240 (2002)

B. Fagan (ed.), *The Complete Ice Age: How Climate Change Shaped the World*, Thames and Hudson, p. 240 (2009)

C. H. Langmuir and W. Broecker, *How to Build a Habitable Planet: The Story of Earth from the Big Bang to Humankind*, Princeton University Press, revised and expanded edition, p. 720 (2012)

J. J. Lowe and M. Walker, *Reconstructing Quaternary Environments*, Prentice Hall, 2nd edition, p. 472 (1997)

W. F. Ruddiman, *Earth's Climate: Past and Future*, W. H. Freeman, 2nd edition, p. 480 (2007)

R. C. L. Wilson, S. A. Drury, and J. L. Chapman, *The Great Ice Age: Climate Change and Life*, Routledge (2003)

J. Zalasiewicz and M. Williams, *The Goldilocks Planet: The 4 Billion Year Story of Earth's Climate*, Oxford University Press, p. 336 (2012)

Future climate change

A. Costello et al., 'Managing the Health Effects of Climate Change', *Lancet*, 373: 1693–733 (2009)

IPCC (Intergovernmental Panel on Climate Change), 'Climate Change 2007: The Physical Science Basis', Contribution of Working Group I to the Fourth Assessment Report of the Intergovernmental Panel on Climate Change, Solomon et al. (eds), Cambridge University Press (2007)

M. A. Maslin and S. Randalls (eds), *Future Climate Change: Critical Concepts in the Environment* (Routledge Major Work Collection: 4 volumes containing reproductions of 85 of the most important papers published in *Climate Change*) p. 1600 (2012)

Mark Maslin, *Global Warming: A Very Short Introduction*, Oxford University Press, second edition, p. 192 (2008)

W. F. Ruddiman, *Plows, Plagues, and Petroleum: How Humans Took Control of Climate*, Princeton University Press, new edition, p. 240 (2010)

N. Stern, *The Economics of Climate Change: The Stern Review*, Cambridge University Press, p. 692 (2007)

G. Walker and D. King, *The Hot Topic*, Bloomsbury, p. 309 (2008)

Fixing climate

R. Gelbspan, *Boiling Point*, Basic Books, p. 254 (2005)

Mark Henderson, *The Geek Manifesto: Why Science Matters*, Bantam Press (2012)

M. Hillman, *How We Can Save the Planet*, Penguin Books (2004)

R. Kunzig and W. Broecker, *Fixing Climate*, GreenProfile, in association with Sort of Books, p. 288 (2008)

C. Hamilton, *Earthmasters: The Dawn of the Age of Climate Engineering*, Yale University Press, p. 247 (2013)

M. A. Maslin and J. Scott, 'Carbon Trading Needs a Multi-Level Approach?' *Nature*, 475: 445–7 (2011)

Climate

A. Meyer, *Contraction and Convergence: The Global Solution to Climate Change*, Green Books (2000)

The Royal Society, 'Geoengineering the Climate: Science, Governance and Uncertainty', The Royal Society Science Policy Centre report 10/09, The Royal Society, p. 81 (2009)

General reading

D. Brownlee and P. Ward, *The Life and Death of Planet Earth: How Science Can Predict the Ultimate Fate of Our World*, Piatkus, p. 256 (2007)

J. D. Cox, *Weather for Dummies*, first edition, p. 384 (2000)

R. Hamblyn, *Extraordinary Weather: Wonders of the Atmosphere from Dust Storms to Lightning Strikes*, David & Charles Publishers, p. 144 (2012)

J. Martin, *The Meaning of the 21st Century*, Eden Project Books, p. 526 (2007)

The Royal Society, 'People and the Planet', The Royal Society Science Policy Centre report 01/12, The Royal Society, p. 81 (2012)

Fiction inspired by climate

D. Defoe, *The Storm*, Penguin Classics, new edition, p. 272 (2005)

K. Evans, *Funny Weather*, Myriad Editions, p. 95 (2006)

J. Griffiths, *WILD: An Elemental Journey*, Penguin Books (2008)

P. F. Hamilton, *Mindstar Rising*, Pan Books (1993)

S. Junger, *The Perfect Storm: A True Story of Man Against the Sea*, Harper Perennial, reissue edition, p. 240 (2006)

J. McNeil, *The Ice Lovers: A Novel*, McArthur & Company, p. 325 (2009)

K. S. Robinson, *Forty Signs of Rain*, HarperCollins (2004)

"牛津通识读本"已出书目

古典哲学的趣味	福柯	地球
人生的意义	缤纷的语言学	记忆
文学理论入门	达达和超现实主义	法律
大众经济学	佛学概论	中国文学
历史之源	维特根斯坦与哲学	托克维尔
设计，无处不在	科学哲学	休谟
生活中的心理学	印度哲学祛魅	分子
政治的历史与边界	克尔凯郭尔	法国大革命
哲学的思与惑	科学革命	民族主义
资本主义	广告	科幻作品
美国总统制	数学	罗素
海德格尔	叔本华	美国政党与选举
我们时代的伦理学	笛卡尔	美国最高法院
卡夫卡是谁	基督教神学	纪录片
考古学的过去与未来	犹太人与犹太教	大萧条与罗斯福新政
天文学简史	现代日本	领导力
社会学的意识	罗兰·巴特	无神论
康德	马基雅维里	罗马共和国
尼采	全球经济史	美国国会
亚里士多德的世界	进化	民主
西方艺术新论	性存在	英格兰文学
全球化面面观·	量子理论	现代主义
简明逻辑学	牛顿新传	网络
法哲学：价值与事实	国际移民	自闭症
政治哲学与幸福根基	哈贝马斯	德里达
选择理论	医学伦理	浪漫主义
后殖民主义与世界格局	黑格尔	批判理论

德国文学	儿童心理学	电影
戏剧	时装	俄罗斯文学
腐败	现代拉丁美洲文学	古典文学
医事法	卢梭	大数据
癌症	隐私	洛克
植物	电影音乐	幸福
法语文学	抑郁症	免疫系统
微观经济学	传染病	银行学
湖泊	希腊化时代	景观设计学
拜占庭	知识	神圣罗马帝国
司法心理学	环境伦理学	大流行病
发展	美国革命	亚历山大大帝
农业	元素周期表	气候